Dr. Harlio & Amy:

This book proves that New Jersey has spawned many great people. William Carlos Williams celebrated, in his idiom, the beauty & importance of the things he saw around him; I hope I have also done justice to this magnificent state.

Best wishes,
Judyann R. Caracio

NORTHERN
NEW JERSEY

The Delaware River forms the
boundary between western New Jersey and the state of Pennsylvania.
Photo by Sharon Sullivan

Northern New Jersey

Gateway To The World Marketplace

Judyann R. Caracio

Corporate Profiles by Paul Lavenhar
and Robert J. Masiello

Produced In Cooperation with the
Commerce and Industry Association of New Jersey
Windsor Publications, Inc.
Northridge, California

Windsor Publications, Inc.—History Books Division
Managing Editor: Karen Story
Design Director: Alexander D'Anca
Staff for *Northern New Jersey*
Manuscript Editor: Lane A. Powell
Photo Editor: Loren Prostano
Editor, Corporate Biographies: Judith Hunter
Production Editor, Corporate Biographies: Una FitzSimons
Senior Proofreader: Susan J. Muhler
Editorial Assistants: Didier Beauvoir, Thelma Fleischer, Kim Kievman, Michael Nugwynne, Kathy B. Peyser, Pat Pittman, Jeffrey Reeves, Theresa Solis
Sales Representative, Corporate Biographies: Rob Ottenheimer
Layout Artist, Corporate Biographies: Angela Ortiz
Designer: Thomas Prager
Library of Congress Cataloging-in-Publication Data: Caracio, Judyann R. Northern New Jersey : gateway to the world marketplace / Judyann R. Caracio ; Partners in progress by Paul Lavenhar and Robert J. Masiello ; produced in cooperation with the Commerce and Industry Association of Northern New Jersey. p. cm.
Bibliography: p.
Includes index.
ISBN: 0-89781-280-8
1. New Jersey – Economic conditions.
2. New Jersey – Economic conditions – Pictorial works. I. Lavenhar, Paul. II. Masiello, Robert J. III. Commerce and Industry Association of Northern New Jersey. IV. Title
HC107.N5C37 1988
330.9749'2034 – dc19 88-17274 CIP

Windsor Publications, Inc.
Elliot Martin, Chairman of the Board
James L. Fish III, Chief Operating Officer

A glimmer of winter light casts a rosy glow above this Sussex County farm. Photo by Michael Spozarsky

This book is dedicated to my mother, Mamie Saveri Caracio, for the help, confidence, and love she has given me. I could not have done it without her.

This charming, scenic mill is located in Hunterdon County. Photo by Michael Spozarsky

Contents

Contemporary commercial spaces contrast with the older buildings of downtown Newark. Photo by Michael Spozarsky

The Commerce and Industry Association published this volume for one principal reason and that is to showcase New Jersey and northern New Jersey in particular in the favorable light it clearly deserves.

New Jersey suffers a perception problem. Visitors and transferring executives often times think of New Jersey as a stretch of turnpike or a chemical forest of tank farms.

The truth is far different. Our region of the state is in reality one of the finest areas in America in which to live or work. Because it is within one of our country's most desirable areas, it is priced accordingly but offers the amenities that any highly desirable area does in each facet of life, including job opportunities, fine residential areas, appreciation of property values, and the quality of life in general.

We know that this prestige book will find many purposes and will go a long way toward clearing up misperceptions. We're confident you'll find it informative and interesting.

Professional business offices will have this exciting "coffee table" book in their waiting rooms, and many private homes will welcome it as an attractive and informative reference book for their personal libraries. Foreign-owned firms will use it constantly for incoming executives. Libraries will find it to be in demand for reference, and schools that offer social study courses will want to include it as source material for special studies.

There have been numerous publications on individual communities over the years, but there never has been one complete volume that embraces northern New Jersey as one great marketing region. This book brings home that point and provides the reader with a broad view of our region and its advantages. This area is second to none in America from any standpoint, and we believe this story has to be told again and again.

New Jersey has been aptly named the Garden State. This volume will help you to understand why.

Jim Cowen
President, Commerce and
Industry Association of
New Jersey

Foreword

This book was difficult to write, primarily because it is hard to keep pace with the rapidly changing terrain of contemporary events. Unlike a history, this book has attempted to capture completely a moment in time when northern New Jersey rides the crest of a tidal wave of unparalleled economic success. If I have been at all successful in conveying the facts plus the energy and enthusiasm to be found everywhere in this region, it is due at least in part to the help I have had. The labor has been ameliorated by many, proving that nothing, including this book, is ever created in a vacuum. My thanks and appreciation must go first to Jim Cowen, president of the Commerce and Industry Association of New Jersey. His readiness to respond to my inquiries made the research task far less laborious than it might otherwise have been. Chip Hallock, executive vice president of the Association, has been similarly helpful by offering background information about the region. Without the Association's efforts, this book would have remained a dream. I am also grateful to my editor at Windsor, Lane Powell, for his sagacious opinions and warm support throughout this project.

My thanks must also go to Elizabeth Harvey, a trusted friend whose intellectual acumen and editorial skills helped to give form and substance to my jottings. Her help was invaluable. Many thanks to Victoria Hardy, former executive director of the William Carlos Williams Center. I would also like to thank Charles Cummings at the Newark Public Library, who offered kind words of encouragement at the outset of this task. It would be impossible to overlook the help given by Carol Duffy at *The Record*. Her immediate and cheerful responses to my smallest request is appreciated. John Zalarick, former director of Marketing Communications at *The Record*, made the newspaper's vast research resources available to me. I would also like to thank Bob Burnett of the New Jersey Historical Society, who gave Chapter One a critical reading.

I am also indebted to John O'Leary of Grant Thornton; James Prior, assistant vice president/executive editor of *New Jersey Business;* Ken Marchi of PSE&G's Area Development department; Frances J. Mertz, president of the Association of Independent Colleges and Universities in New Jersey; Chancellor T. Edward Hollander and the New Jersey Department of Higher Education; Ruth Van Wagner, director of the Office of Cultural and Historic Affairs in Bergen County; Sarah Fryberger, assistant editor of *New Jersey Monthly;* and Cynthia Wilk, development director of the Garden State Ballet.

I would also like to thank all the librarians who have been there to offer assistance; in particular, Evelyn Klingler of the Morris County Library. Her extensive knowledge and helpfulness saved me many hours of frustration. New Jersey Hospital Association's fine medical library made it possible for me to address the complicated topic of health care in New Jersey.

I would also like to thank John T. Cunningham for writing so many interesting and informative books about New Jersey. Their existence made my job easier.

I couldn't have accomplished this job without the piles of books, magazines, newspapers, brochures, etc., that provided me with much of the background. I am deeply grateful to all of the agencies, organizations, and businesses that provided them.

Above all else, a special thanks from my heart to R.F., who listened with great patience over many months to daily reports on the progress of "the book." The love and support he offered during the difficult times can never be repaid. His intellectual contributions permeate these pages, and were responsible for shaping this manuscript *ab ovo* to completion.

Acknowledgments

The historic railroad building of Liberty State Park remains illuminated throughout the evening. Photo by Mark E. Gibson

AND YET THEY COME …

Hackensack appears to be imbued with the spirit of improvement which seems to know no cessation. There are now perhaps forty new buildings being erected. Our village has doubled its population within a few years and yet they come. We have all the comforts and appliances of city life, with the advantage of cheaper living …Jersey being the only outlet for the superabundant population of the city, the attraction of our village will induce hundreds of others, wealthy and discriminating citizens, to settle in our midst, and thus enjoy our advantages and contiguity to the city.

Bergen County Democrat, *November 13, 1868*

…**A**nd yet they come to northern New Jersey as they have from the beginning, when its beautiful shoreline beckoned to the first Europeans who sought the riches of the new world. Since that distant time, the region has more than fulfilled its potential. In each "paradigm shift" northern New Jersey has had "the right stuff" to compete and prosper. Its enormous energy resources facilitated an easy transition from an agrarian society to the industrial age. Now, its endless storehouse of talented and educated people is making the segue into an information age an easy task. Northern New Jersey's successful metamorphosis has served as a model for other parts of the country as well. During the past several years the region's economic prosperity has been a source of national envy and local pride.

The area's fortunate geography has helped to nurture its phenomenal growth. At the geographic center of the world's richest market, northern New Jersey is a natural location for both American and international corporations. Nestled in idyllic campus settings throughout the region, numerous companies can still enjoy proximity to New York City. Poised as the heart of a transportation system that stretches out by highway, rail, ocean, and air, northern New Jersey puts people and products within easy reach of suppliers and consumers anywhere in the nation, and the world, as well. Clearly, northern New Jersey is the Gateway To The World Marketplace.

Introduction

*An idyllic farmhouse on Sunrise
Mountain is evidence of northern
New Jersey's agrarian beginnings.
Photo by Michael Spozarsky*

Long before European eyes beheld the natural beauty and abundance
of New Jersey, the aboriginal Indians owned the land. Their name for it
was "Sagorigiviyogstha": the doer of justice. The rich, fertile soil
waited for centuries before the white man—the new immigrant—trod
upon it. The native progenitors were known as the Lenni Lenape Indians, or "Grandfather Tribe." This important branch of the
Algonquin family had laid claim to New Jersey on their journey westward. The tribe's long habitation in New Jersey gives a poignant
significance to the name Lenni Lenape: "Lenni" means pure or original, and "Lenape" means people. Their name, then, may mean "the
first people." However, it was only

CHAPTER
ONE

a matter of time before these origi-
nal inhabitants had to share the wealth of their land with newcomers.

DISCOVERY AND SETTLEMENT   That time finally arrived
when the great maritime nations of Europe searched for shorter trade
routes to the Orient. Each nation "discovered" New Jersey in its quest
for the Northwest Passage. John Cabot (Giovanni Caboto), an Italian
explorer living in Bristol, England, was the first. In the employ of King

Building On A Tradition Of Leadership

The rich farmlands of northern New Jersey are owned and operated by resident farmers. Photo by Michael Spozarsky

Henry VII of England, Cabot and his son Sebastian explored the eastern coast of North America in 1498. Sailing from Labrador to Florida, the Cabots discovered New Jersey's coastline. They were the first Europeans to view the land, yet they never disembarked from their ships. They made no attempt to settle the land and establish a legal right to it; they simply declared it the property of the Crown of England. Cabot's discovery gave England its claim to the eastern coast of North America and prepared the way for the founding of the English colonies in the New World.

After Cabot's sighting of New Jersey, Giovanni da Verrazano, an Italian navigator in service to Francis I of France, sailed along the eastern coast of North America in 1524. He declared it "a new land never before seen by any man, ancient or modern," according to Frank R. Stockton in his book *Stories Of New Jersey*. Verrazano took possession of the land in the name of his king. To assure the legitimacy of the country's right of ownership, he named it New France.

In 1609, Europeans again visited New Jersey when the Englishman Henry Hudson, under the auspices of the Dutch East India Company, sought a northwest passage to China. Sighting New Jersey from the south, he partially explored Delaware Bay before sailing the *Half Moon* up the river that now bears his name. He and his men were the first Europeans to set foot on the soil, some of them landing in the vicinity of Bergen Point. But it was not until the mid-1620s that the Dutch returned to exploit Hudson's claim.

The native Americans who occupied the area were led by Oratam, the great sagamore and sachem of the Hackensack (Achkinkeshacky)

Bales of hay drying in the fields of Sussex County are soon to be collected. Photo by Michael Spozarsky

The Dey Mansion in Totowa is one example of early Dutch architecture in northern New Jersey. Photo by Sharon Sullivan

Indians. Oratam's intention to pursue a peaceful coexistence with the new arrivals allowed European settlement to occur more quickly than it would have otherwise. Evidence of his attempts to resolve problems appears in treaty after treaty bearing his name. The dogged perseverance of the Dutch established the white man in the more northern reaches of the state. At first, the Dutch interest lay only in the land's natural bounty that could be gleaned and sent back to Holland. By the mid-1600s, however, they became interested in more than trading. They began to think of the New World as a place to put down roots rather than as a mercantile outpost in a primitive territory.

In 1630, the first permanent settlement in New Jersey was established directly across the Hudson River from the small Dutch fortress at New Amsterdam. Dutch Burgomaster Michael Reynierse Paauw made a pact with his government to deliver 50 settlers to the area that now encompasses Hoboken and Jersey City, which were part of Bergen

County until 1840. Since the name Paauw means "Peacock" in Dutch, he named the area Pavonia—"The land of the Peacock." John Cunningham notes in his book *This Is New Jersey* that Paauw failed to fulfill his promise to "plant a colony of fifty souls, upwards of fifteen years old." However, Pavonia remained in existence as a valuable site for traders.

Two of the earliest settlements in what is now Bergen County were established in 1640 but were destroyed a few years later. Both David DeVries' outpost at Vriessendael, which included the present borough of Edgewater, and Achter Col on the east bank of the Hackensack River in modern Bogota, were destroyed in Indian attacks. The Dutch lack of regard for Indian rights inevitably led to confrontations. William Kieft, a Dutch governor, exhibited gross inhumanity when he ordered the massacre of 80 friendly natives. This precipitated years of Indian retaliation.

In 1660, Governor Peter Stuyvesant approved the establishment of New Jersey's first town: Bergen. It was built behind a square wooden barricade that protected against Indian attacks. The village of Bergen was located near today's Journal Square in the heart of Jersey City.

most of present day Clifton, Passaic, and Paterson. In fact, the Dutch had some influence in both Passaic and Hudson counties because they were both part of the original tract of land known as Bergen County. Even today, distinctive Dutch architecture dots Bergen County's landscape. Although there are no examples of the earliest Dutch Reformed churches (remaining ones date from the 1790s), homes with thick stone walls and strong gambrel roofs offer evidence of Dutch history. The Dutch people's language, architecture, and culture gave Bergen County a patina unlike that of anywhere else in colonial America. The county's history remains singularly intriguing to those interested in America's ethnic influences.

AGRARIAN BEGINNINGS From the outset, for the Dutch, the land seemed to offer more agreeable terrain than its stony Manhattan counterpart. The Dutch colonists found the luxuriously rich soil they required for farming behind the grand rocky facade of the Palisades, which stood like formidable giants protecting the natural riches they concealed. The Jersey Dutch came to possess the very garden spot of America and were envied by colonists of every province. Farmers flourished as a result of their initiative and ingenuity, traits that were to become commonplace attributes of northern New Jersey residents. Their enterprising spirit transformed agriculture from subsistence farming to a commercial enterprise. By the 1770s, the prosperity of this area was well known in what was by then British America. If one were to believe the fantastic accounts of the day, the soil yielded produce of gargantuan proportions: 100-pound pumpkins, one-pound pears, and celery more than a yard long. By 1859, the strawberry trade had won national acclaim, and millions of baskets were produced annually. Ramsey and Allendale became centers for the industry. The fruit had to be sent in wagonloads over rough roads to market; this stimulated farmers to petition for improved roads and bridges to speed the journey. Bergen County was solidly on its way to becoming a healthy agrarian society.

The Bank House built in 1831 by the Washington Banking Co. was the first bank in Bergen County. This Federal style building was enlarged in 1909. Photo by Carol Kitman

Local strawberry farms invite the public to "pick their own" berries. Photo by Michael Spozarsky

THE DUTCH INFLUENCE Slowly, a unique Dutch cultural flavor began to surface. With the exception of Essex County, which was settled by the English under Robert Treat, the Dutch were instrumental in settling many areas of northern New Jersey. In 1695, Dutch entrepreneurs came from Bergen to settle 1,500 acres near Pompton Plains in Morris County. Dutch wanderers trekking overland from Philadelphia to New York stopped at Schooley Mountain, also in Morris County, in 1707. In Passaic County, the Dutch came early to settle in a spot below the "Great Falls." But the earliest settlement in Passaic County was established by Hartman Michelsen, who purchased an island in the river in 1679 from Captehan Peeters, an Indian. In 1682, 14 Dutch families set down roots in the Acquackanonck Tract that comprises

"Driving along a country road in northwest New Jersey, we ran across a magnificent farm." Photo by Michael Spozarsky

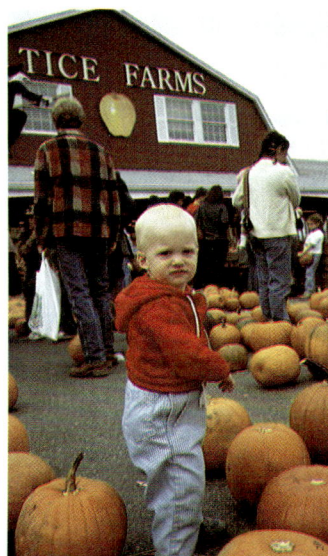

This young man selects his Halloween pumpkin from Tice Farms in Woodcliff Lake. Photo by Carol Kitman

OTHER EARLY COMMERCIAL ENDEAVORS   The county's natural resources provided the raw materials for other commercial endeavors that, although they did not eclipse the strawberry trade, were harbingers of things to come. The Bergen County iron industry flourished in the area that is now Passaic County. Copper was mined in the Schuyler Copper Mines found within present-day Bergen County. Mention of the mines are found as early as 1715, and they were worked for two centuries before flooding plagued owners and caused a decline in production. Lumbering was also a colonial endeavor, but was destined to meet its demise as forests rapidly dwindled. Clay found along the Hackensack River accommodated the formation of brick and pottery works late in the nineteenth century. The importance of waterpower was recognized early on, and grist- and sawmills sprang up along the county's many waterways. Later, waterpower led to industries like textile mills and iron foundries, which would facilitate the area's easy transition into the industrial age.

One of the most interesting and ingenious industries to surface at the time sprang from one man's ability to perceive a need and to fill it. As Indian traders searched for the means to conduct business with native Americans, they found that John Campbell and his sons could fulfill their demands for wampum, Indian "currency." By 1775 Campbell was manufacturing strings of drilled and polished shells in the volumes needed for trade. The cottage industry grew to occupy a factory-sized building. Conch and clam shells collected from New York City fish markets were transformed into wampum by farmers' skilled wives and daughters. Women produced five to ten strings daily for 12 and one-half cents per string. The process was quickened when two of Campbell's sons invented a drilling machine that produced wampum "pipes," or the cylindrical ornaments Indians wore on garments or in their hair. Campbell's splendid example of northern New Jersey entrepreneurship helped open the American West. Traders looked to his fac-

tory on the Pascack Creek for the trade goods they needed. Even such famous adventurers as John Jacob Astor, who needed wampum to conduct his fur-trading ventures in the West, came to Campbell. The colorful industry went out of fashion in the late 1800s when "Manifest Destiny" had been secured and wampum was no longer required. But by that time, John Campbell and his sons had made their contribution to the nation's growth as well as to Bergen County's history.

INDUSTRIAL BEGINNINGS Within the framework of an agrarian society, northern New Jersey was by the late seventeenth century already sowing the seed for the great Industrial Age to come. Well before the Revolutionary War, Lenni Lenape braves showed the white men "Sucky sunny" or "black stone," referring to the Anglo name for iron. Much of Morris County's story begins with the discovery of this iron. By 1710, the old forge on the Whippany River made Hanover an

The Phoenix Mill is located in the Paterson Historic District. Photo by Michael Spozarsky

General George Washington maintained his headquarters in Morristown for two winters during the American Revolution. Courtesy, Cirker, Dictionary of American Portraits

eral George Washington was well aware of the importance of Morristown's industry. In fact, the weapons needed to supply the Continental Army came in a steady stream from Morris County forges and furnaces. Washington was also aware of Morristown's easily defensible geography. During his first Morristown winter in 1777, he maintained his headquarters in Jacob Arnold's Tavern; his second in 1779 was spent in the Ford Mansion. Clearly, Morristown played a dual role as munitions maker and innkeeper for the American army under siege. The iron industry continued after the war, and by 1880 it had grown enough to make Morris County third in the nation in iron ore production.

After the war, in the beginning of the nineteenth century, the ingenuity and vision of George P. Maculloch, a Morris county schoolteacher, led to the construction of the Morris Canal, which linked Pennsylvania's anthracite coal fields with northern New Jersey's ironworks. By the mid-1800s, the mountain-climbing canal brought vessels from Easton, Pennsylvania, to Newark. The canal was a great boon, providing cheap transportation and alternative fuel to timber for the forge fires. However, the canal's construction could not save the county's moribund mines; it could only provide a temporary revival of the former prosperity. But that reality does not lessen the enormity of Maculloch's imaginative powers and dedication. He proved that what can be conceived can be created. Maculloch exemplified the ingenuity and vision typical of northern New Jerseyans.

Iron was also an early factor in Passaic County's growth before the establishment of Paterson's industrial might. Under the direction of the American Company (sometimes called the London Company), Peter Hasenclever came from Germany in 1764 to establish an iron industry. With the help of 535 German immigrants, he fashioned a thriving community at Ringwood. This industrial spurt hastened road building which connected the mines to furnaces in surrounding towns. Like Morristown's iron industry, Ringwood's furnaces provided munitions for the American Revolution.

important town. Shortly after, Morristown gained fame for this forge, and within 20 years there were ironworks in Dover and Rockaway. Centered near rivers, the industry thrived on the easy availability of waterpower and timber used for forge fires. Most of the ore came from the mine near Succasuna, where "black stone" could be found on the surface with little excavation. By 1750, none could disagree that Morris County had become the center for ironworks. The industry had become so entrenched that county ironmasters were willing to disobey a decree by the British that forbade rolling and slitting mills in the colonies. John T. Cunningham asserts that patriotism was not the motivating force behind the colonists' behavior. Instead, he claims that "this rebellion of the 1750s was more a rebellion of the pocketbook than of principle."

When the war for America's independence was fully under way, Morristown's iron industry was placed solidly behind the cause. Gen-

Alexander Hamilton, a political leader, and a personal advisor to George Washington, was responsible for the ratification of the U.S. Constitution. Courtesy, Dictionary of American Portraits

The Great Falls of Paterson provide a stunning backdrop for this tightrope teaser. Photo by Michael Spozarsky

The Indians were the first to understand the might of the "Great Falls" at Paterson, which were to play a vital part in northern New Jersey's transition into the Industrial Age. They called it Totowa: "heavy falling weight of waters." However, it was Alexander Hamilton who understood the falls' full import when he stopped at their foot for lunch with Washington in 1778. Thirteen years later, when Hamilton was secretary of the treasury, he sent to Congress his "Report on Manufactures." He reasoned that a nation could never be free until it manufactured its own products. On November 22, 1791, Governor William Paterson signed the charter for the Society for Establishing Useful Manufactures, and Paterson was launched as the first planned industrial city in America, becoming the archetype for similar cities that followed. With energy as the most important necessary component, Paterson, with its raging falls, was at a distinct advantage. Its illustrious history as a center for manufacture was assured.

With the contributions of countless creative northern New Jerseyans, Paterson set the tone for the coming industrial era. John Parke built a small cotton mill, which he expanded in 1810. Others were built, and soon several cotton mills were producing efficiently. In 1827, John Colt was the first to substitute cotton for flax in sail duck. His yacht, *America,* won for this country the coveted America's Cup. Thomas Rogers founded the Rogers Locomotive Erecting Shop in 1871. At the height of production, a locomotive rolled out of his factory every two days. During the next 50 years, five other locomotive firms were established in the city. They produced many of the engines that were instrumental in settling the West. Another innovator, Samuel Colt, produced his now famous Colt revolver in Paterson in 1836. With Rogers' steam locomotives to carry the pioneers, and Colt's guns to arm them, Paterson's best had made the West more accessible and safer. Years later, in the 1920s, the engine for Charles Lindbergh's *Spirit of St. Louis,* manufactured at Wright Aeronautical Corporation, made the world more accessible by changing forever the means of travel. Paterson, then, had been instrumental in expanding the nation's and the world's possibilities. In the early 1840s, silk manufacture was introduced by John Ryle, an Englishman, and Paterson came to be known as the "Silk City." At one time the city produced more silk products than any other American city. As Paterson prospered, so did the nation. And New Jersey, on the whole, took her place among the great manufacturing states of the Union.

Newark's fame also grew, and it took a spot next to Paterson as a hub of industrial activity. Its remarkable ascendancy to the status of a major industrial center was a matter of local pride. By the first quarter of the twentieth century, it could boast that no other town manufactured a greater variety of products. Today, with the proximity of major highways, an international airport, and a seaport, Newark retains its reputation as a locus of local, national, and international transportation activity.

FURTHER REVOLUTIONARY CHANGES By the first half of the nineteenth century, northern New Jersey had undergone a metamorphosis. Her lively agrarian underpinnings remained, but now they

Rogers Locomotive Museum and Erecting Shop are popular sites in the Great Falls Historic District of Paterson. Photo by Michael Spozarsky

The Rogers Locomotive administration offices are located here in the Silk Machinery Exchange Building of the Great Falls Historic District, Paterson. Photo by Michael Spozarsky

were joined by a growing industrial might. By the close of the Civil War, northern New Jersey, as well as the rest of the state, was revolutionized by industrial expansion. Cities burgeoned, population poured in from New York and Philadelphia, and the great captains of industry made their mark. Material ideals were paramount as men made vast fortunes.

The transportation revolution in the last half of the nineteenth century brought to fruition the expansion and economic surge foreseen as early as the Civil War by savvy real estate entrepreneurs. Population followed the rail lines, which were completed by the early 1880s. In Bergen County, real estate agents enticed Manhattanites to cross the river with the promise of easy access to their city jobs by train. "Land Auctions" helped create the county's many separate and autonomous municipalities, which sprang up along railroad routes. The modern commuter was born. To accommodate the commuter, a plethora of services were created, and neophyte communities grew into sophisticated suburban enclaves.

With the advent of a new century, northern New Jersey was poised to take advantage of an infinite array of possibilities. As Catherine M. Fogarty, John E. O'Connor, and Charles F. Cummings note in *Bergen County: A Pictorial History*: "The transformation of values was spurred on by the cultural entrepreneurs, the magazine and newspaper editors, the lecturers and artists and others involved in the commercialization of American culture."

Railroad tracks were being laid in New Jersey as early as 1825 to accommodate either a horse-pulled or a steam-powered locomotive. Photo by Michael Spozarsky

*The George Washington Bridge
proudly spans the Hudson River
connecting Fort Lee, New Jersey,
with the "Big Apple." Photo by Bob
Krist*

One of the most flamboyant commercial manifestations of the era was born in West Orange when Thomas A. Edison experimented with the first motion picture camera in 1888. Edison's film "The Great Train Robbery" was the first to have a plot. It was filmed at the Lackawanna Railroad station in Paterson. However, as the dramatic scope of the films widened, there was a need for broader, more realistic settings. The novice filmmakers found what they wanted atop the Palisades in Fort Lee. The town, which had once seen the drama of the American Revolution unfold, was now host to celluloid dramas. The semi-rural town of Fort Lee and the rocky bluffs of northern New Jersey soon became the Wild West. D.W. Griffith, Mack Sennett, William Fox, Samuel Goldfish (later Goldwyn), and Jesse Lasky all got their starts there. In 1909, the Champion Film Company became the first studio. Eclair American Company followed, employing the likes of Mary Pickford, Lionel Barrymore, and Lillian Gish. The Victor Film Company, Solax Company, and World Pictures also had headquarters in Fort Lee.

But by 1924 Fort Lee's glory days of motion picture production were spent. World War I had caused a shortage in coal, and operations were moved to the warmer West. Most of the studios were later absorbed into larger companies, creating the studios we know today: Paramount, Universal, MGM, etc. However, northern New Jersey had played an important part in shaping commercial mass entertainment.

Edison's contribution to the motion picture industry represents a miniscule portion of his prolific output as an inventor. His genius literally electrified the world from his "invention factory" in Menlo Park, which was the first of its kind. From the two-story clapboard building, the "Wizard" lit up the world with his incandescent light bulb. Ten years later, he moved his lab to West Orange. His assemblage of mathematicians, chemists, modelmakers, mechanics, and amateur tinkerers produced a staggering number of inventions. Edison had more than a thousand patents to his name, but his greatest invention remains unpatentable: the first organized research and development laboratory.

Edison may be northern New Jersey's most famous resident inventor, but he is by no means its only contender for the title of inventive genius. It is only fitting that the state that "invented" the research lab should have an inexhaustible pool of inventive talent. Today, the vast numbers of scientists and engineers living in the state give it the highest per capita number of resident scientists in the nation—a fact that tempts many to dub New Jersey the "Research State."

The northern New Jersey tradition of invention has manifested itself throughout the area's history. Its many tinkerers have, in many cases, changed the world. John Stevens ran the first American steam locomotive at Castle Point in Hoboken in 1825. The Hyatt brothers perfected celluloid in Newark. Alfred Vail, working with Samuel Morse, transmitted the first long-distance telegraphic message, from Speedwell Iron Works in Morristown. John Holland designed and built the first submarine in Paterson in 1878. Seth Boyden, tinkerer extraordinaire of Newark, invented patent leather and found the secret

*Thomas Alva Edison records his
voice on the wax cylinder of an
Ediphone, which was the precursor
of the phonograph record. Courtesy,
Newark Public Library*

of malleable iron, among countless other discoveries. Anthony Fokker, the Dutch aircraft designer and manufacturer, paved the way for commercial aviation in the United States with his tri-motors, which were assembled at Teterboro. Northern New Jersey's abundant supply of people endowed with the "spirit of enterprise" has never diminished. Each wave of newcomers brought a fresh supply to her shores. The contributions of the varied ethnic groups added a rich texture to life in the northern communities.

One of the greatest catalysts in northern New Jersey's recent history for the influx of talented new residents was the construction of the George Washington Bridge. Developers had heralded the coming of the bridge for years before its actual construction. Charles Reis had created "Sunshine City" in Wood-Ridge in the late 1920s as an early attempt at mass-produced housing. Radburn, in Fair Lawn, one of the first planned communities, offered residents a self-contained, pedes-

THE COMMERCE AND INDUSTRY ASSOCIATION OF NEW JERSEY

The Commerce and Industry Association of New Jersey offers a meeting ground where business opportunities are nurtured and where common interests are shared. Here association leaders are pictured at a recent gathering.

Northern New Jersey is a major business center, strategically located for domestic and international trade. It is also a prestigious residential area, complemented by outstanding shopping and recreational facilities. The Commerce and Industry Association of New Jersey has sponsored this book to illustrate how it represents one of the finest regions in the nation in which to conduct business, live, and enjoy leisure time.

For more than 60 years the association has served the area with major emphasis on developing and protecting the opportunities for businesses to locate and prosper there. Through its broad membership the association works with all levels of government to maintain a favorable business climate. It advocates a healthy business base as essential to providing area residents with rewarding careers, while also establishing a tax base that ensures development of resources and services to benefit all segments of the state's population.

The association offers a meeting ground where business opportunities are nurtured and where common interests are shared. Its success is demonstrated

The Commerce and Industry Association of New Jersey is now the largest organization of its kind in the state. In this picture the association leadership discusses policy relations.

by the membership of the board of directors, which reflects a cross section of top executives of major corporations locating in increasing numbers in northern New Jersey, as well as firms with deep roots in the region.

The Commerce and Industry Association is now the state's largest organization of its kind. Promoting free enterprise, economic education, and the privatization of government services are among its top priorities.

The association relies heavily on the participation of members through various committees. These groups meet regularly to develop strategies, and to arrange meetings with federal and state legislators and other government officials.

The association's publications program provides an invaluable service to its members. The award-winning monthly magazine *Commerce*, a monthly newsletter, special bulletins, wage and salary studies, and news of major corporate moves and relocations provide a flow of information of great value to the operations of area firms.

The Commerce and Industry Association is a strong voice for free enterprise in the legislative and educational arenas. Through the Private Enterprise Political Action Committee (PENNPAC), financial support is provided to candidates for state office considered to best understand the concepts of the free market and a vigorous economy for the state. The Foundation for Free Enterprise, a Commerce and Industry Association affiliate, provides economic education opportunities for teachers and students at all school levels. Since 1975 extensive information on the operation of the free enterprise system in the marketplace has been furnished to more than 5,000 students and teachers.

In these and other ways, The Commerce and Industry Association of New Jersey serves the business community of northern New Jersey.

Nightfall surrounds the Historic District of Paterson. Photo by Michael Spozarasky

trian community sans the overt intrusion of automobiles. Clearly, even the rumors of a bridge precipitated astounding results. When the bridge finally opened on October 24, 1931, Bergen County braced itself for a surge of humanity. Just as northern New Jersey's strategic location had put it in the direct path of the early explorers of this country, it also made the area the gateway to successive waves of new Americans seeking a better life. The early Dutch "commuter" from New Amsterdam was supplanted by the suburbanite who worked in New York City and lived across the river. Using northern New Jersey as their "bedroom," commuters enjoyed the best of both worlds: proximity to New York's advantages and a quiet life in a suburban setting. However, recent studies have shown that suburbanites no longer consider themselves dependent on New York City for jobs or cultural and social activities. Fewer and fewer go into Manhattan because local communities provide jobs as well as a panorama of social and cultural events. Commuting, once the standard way of life for many, is no longer the rule.

Once the commuters had "discovered" northern New Jersey and its advantages, it was only a matter of time before corporations followed closely in their path. Over the years, northern New Jersey has become a haven for corporations seeking to escape the high costs of doing business in New York City. Other reasons for their choice are legion: the region's ideal location; its proximity to customers and to other company installations; excellent local, regional, national, and international transportation facilities; community receptivity to business and industry; the availability of highly qualified professional, technical, and clerical workers; and a style of living that suits employees at every level. It is clear that northern New Jersey is in the throes of a corporate explosion, and the transition has created abundant job opportunities and a way of life that is surpassed by few places in the nation.

Northern New Jersey's inexhaustible vigor has made it a leader in every societal shift America has made—from agrarian to industrial, from industrial to information. As John Naisbitt points out in *Megatrends,* we are in the middle of an economic evolution, and we are watching the "restructuring of America." Nowhere is this transition more evident than in northern New Jersey. Its ample storehouse of inventive, skilled workers "stoke the fires" of the information society the same way its massive waterpower once generated the impetus to build and maintain its industrial might. Northern New Jersey has had, from the beginning, the right ingredients. Her success is a *fait accompli.*

Major corporations find the economic diversity of northern New Jersey very attractive. Photo by Don Klumpp

Geographically positioned at the center of the greatest marketplace in the world, New Jersey's economic prowess over the past several years has made it the envy of the East. Flanked by two major metropolitan centers, New York and Philadelphia, the state has sculpted its own unique identity as a formidable titan of commerce and industry. In addition, the state's ongoing economic expansion shows no signs of old age, making it the second-longest in postwar history; only that of the Vietnam War era was longer.

However, New Jersey's astounding success is built on more than geography. The diverse nature of New Jersey's economy with its strong base in the service industries has helped the state to avoid the economic problems experienced by states with focuses on energy and agriculture. Because of this, New Jersey has been buoyed to success while other state economies have floundered. The strength of that economy has elevated the state's per capita income to $18,248—the second highest in the country after Connecticut. With the ninth-largest population in the nation (7.7 million), New Jersey still enjoys the second-lowest unemployment rate in the United States, hovering at approximately 3 percent. Over the last four years, over 500,000 jobs

CHAPTER TWO

A Corporate Mecca

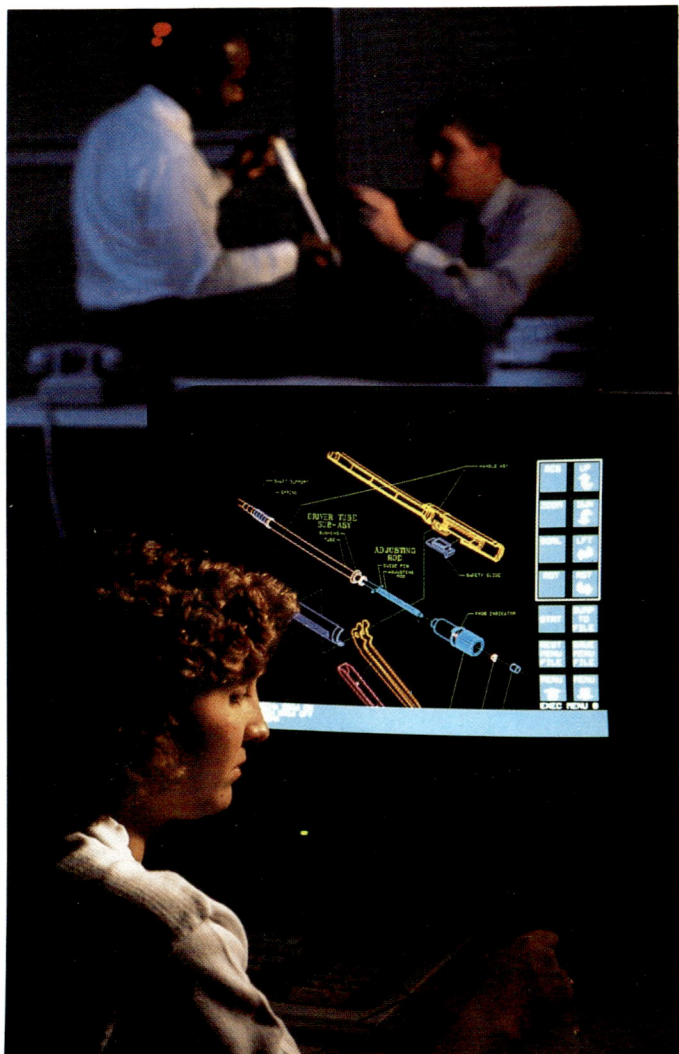

The current economic expansion in New Jersey is boundless. Not only is the per capita income for the state of New Jersey second in the country, the unemployment rate is second lowest in the country. Photo by Bob Krist

have been created by the state's healthy economy. In fact, outperforming the nation has become a way of life in New Jersey. More office buildings are under construction in New Jersey than in any other state, and it outpaces the nation in retail sales. New Jersey also boasts an enviable infrastructure—the busiest airport in terms of takeoffs and landings (Teterboro), the biggest "containerized" shipping port (Port Elizabeth), the most heavily traveled toll road (the New Jersey Turnpike), and the largest private business in the world (AT&T).

These factors, together with the state's pro-business environment and the energy and talent of its well-educated labor pool, have been key factors in making it especially attractive to large corporations seeking a place to put down roots, and many do just that. Although New Jersey is the fourth-smallest state in the Union, it has the third-largest number of corporate headquarters. Drawn by the lure of more space, low operating costs, and an alternative to cumbersome commutes, businesses

are flocking to suburban settings. In addition to companies relocating from other areas of the country and the world, most New York businesses find the Garden State alluring. According to Eugene Heller, president of Hartz Mountain, in an article in an area business publication, "virtually every Manhattan-based corporation considering expansion is taking a hard look at New Jersey alternatives." With a 40-percent savings on such basic costs as electricity and taxes, there is ample motivation for most companies to consider New Jersey.

Northern New Jersey, in particular, has been the destination for many of the largest American and international corporations. Its excellent quality of life and solid professional work force have enticed many of the giants to take up residence in its environs. With northern New Jersey's accelerated business activity, George Sternlieb and Alex Schwartz of Rutgers University's Center for Urban Policy Research have identified four prime high-growth "corridors" in northern New Jersey: northern Bergen County, the Meadowlands, interstates 78 and 287 in Morris County, and the Hudson River Waterfront. Northern Bergen and Morris counties, in particular, with their expanses of verdant landscape, have become sought-after sites for corporate headquarters. Moving from New York to serene campus-like settings in those areas has afforded companies and their personnel an opportunity to become part of their respective communities and to enjoy their work and their surroundings.

Four of New Jersey's best constitute the following representative sampling of the diversity among American businesses residing in northern New Jersey:

J. Fletcher Creamer & Son—As one of the most successful construction companies in the state, J. Fletcher Creamer & Son, Inc. (JFC), has been proud to call New Jersey its home since the company began in 1924. Company president J. Fletcher Creamer is the third generation of the Creamer family to head the northern New Jersey firm. Creamer is an avid New Jersey supporter when he talks about its advantages:

I wouldn't live anywhere else; I love New Jersey. This state offers everything, as far as I'm concerned: the mountains, the shoreline, open spaces and farms. The proximity to the biggest city in the world is a plus—you have all the benefits of New York City and none of the hassles. I lived in Fort Lee all my life, until ten years ago, and I could leave my house in the evening and be sitting in Madison Square Garden in twenty minutes. However, the inconveniences of city living do not have to be part of my everyday existence.

As a leader among New Jersey firms that have diversified within the construction marketplace, JFC has become one of the most successful contractors in the northeast. Working primarily in the utility field until the 1960s, recently JFC has been building structures. "We have been building bridges," Creamer said, "and now we have expanded on that, and we are doing more highway work."

JFC's recent activities are a far cry from its work during its early years in Fort Lee, in 1924, when the humble company owned only one

The Schering-Plough Corporation produces pharmaceuticals in Kenilworth. Photo by Bob Krist

COLE, SCHOTZ, BERNSTEIN, MEISEL & FORMAN, P.A.

Michael S. Meisel and Morrill J. Cole

I. Garth, was also appointed to the United District Court for the District of New Jersey and now serves on the United States Court of Appeals for the Third Circuit.

These appointments underscore the character and superior abilities valued by the firm's partners and associates alike. In addition, a large segment of the firm's current partners and associates have been published in major law reviews and periodicals, lecture regularly before professional and industry associations, and serve as faculty members at law schools within the New York metropolitan area.

In 1982 the Cole firm merged with Shavick, Schotz, Nadler & Konner, which also had its roots in America's first industrial city, Paterson. Together they form the current organization, which is headquartered in the Court Plaza North complex across from the Bergen County Courthouse in Hackensack.

While the firm's roots extend back some six decades of this century, its attitudes and approaches to the practice of law are highly contemporary—without losing CSBM&F's commitment to a tradition of service. The firm—the largest in Bergen and Passaic counties—has organized itself along specialized areas of concern, such as litigation, taxation, real estate, commercial, and corporate issues.

Cole, Schotz, Bernstein, Meisel & Forman offers a complete range of legal services. It counsels in the formation of new businesses, the ongoing conduct of businesses, and in the modifications and adaptations required by an ever-growing body of regulatory laws. Its litigation department handles commercial cases in the state and federal courts involving complex contract, securities, unfair competition, and accounting issues.

CSBM&F's tax and estate department offers tax planning covering business, estate, and personal issues. The real estate department conducts all aspects of commercial, industrial, and

The astounding economic growth and increased sophistication of the northern New Jersey region can be attributed to many factors. These include its location at the outskirts of the nation's largest city and financial center, a core of entrepreneurial individuals and business enterprises willing to invest in the area, superior transportation facilities and access, a large capital base, and other variables.

One important element has been the level of legal services and expertise offered to local entrepreneurs and major corporations locating within the region. Today it is not necessary for New Jersey businesses to cross the Hudson River to find the highest caliber law firms to serve their needs and interests. The concentrated abilities and tenacity long associated with the proverbial Wall Street law firms can be found in northern New Jersey's finest practices, such as Hackensack-based Cole, Schotz, Bernstein, Meisel & Forman, P.A.

Dating from 1928, when its first antecedent, Cole & Morrill, was established in Paterson, the firm has set an example for combining private and public service. David L. Cole, labor arbitrator and mediator, served as director of the Federal Mediation and Conciliation Service, and counseled every president from Harry Truman to Gerald Ford in national labor crises. Mendon Morrill, one of New Jersey's foremost trial attorneys, culminated his career by serving as a judge on the United States District Court for the District of New Jersey. Another former partner, Leonard

Michael H. Forman

residential transactions, while the firm's business planners and litigators also deal with insolvency matters, including the reorganization of business and the representation of creditors and debtors.

Owing to the multifaceted requirements of its corporate and institutional clientele, CSBM&F carries its services beyond its individual areas of specialization by developing a coordinated approach to clients' needs. The synergy growing out of the firm's particular areas of expertise gives clients capabilities that are truly greater than the sum of their parts.

The broad range of the firm's client base clearly demonstrates its ability to provide a full spectrum of legal services. Cole, Schotz, Bernstein, Meisel & Forman represents the gamut from accounting firms to textile companies, including financial institutions, commercial and industrial real estate developers, environmental firms, manufacturers, professional sports and entertainment interests, medical and dental practices, import-export businesses, high-technology companies, and many other categories of private and public corporations requiring superior legal assistance.

Service, experience, and innovation

have placed Cole, Schotz, Bernstein, Meisel & Forman at the vanguard of law firms serving the northern New Jersey market. From such projects as the development of the Glenpointe multi-use complex in Teaneck to the precedent-setting broadening of the powers of savings and loan associations, which

helped revolutionize the financial services industries, CSBM&F's scope of work has grown over the years.

Cole, Schotz, Bernstein, Meisel & Forman's commitment to skilled and attentive legal services for its clients remains constant today and for the future.

Edward M. Schotz

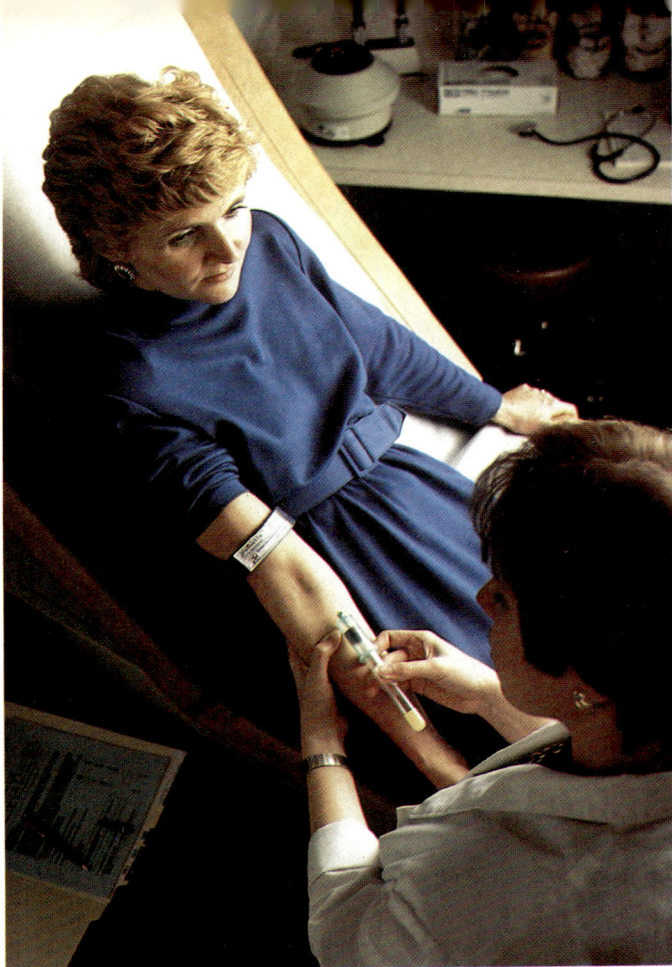

truck. Today, the firm has an Astar Mark III helicopter, which saves the company time by transporting key personnel from one site to another. Utilizing five regional offices—three in New Jersey and two in New York—the firm employs 500 people who carry out diverse and sophisticated projects. For example, the firm recently became involved in the fiber optic market. In the area of heavy construction, JFC constructed the first phase of the development of Port Liberté along the Hudson waterfront in Jersey City. The company has also been responsible for improvements of service areas along the Garden State Parkway, most notably in Cheesequake Service Area and in Montvale. In a $9.5 million contract, the biggest highway job to date, JFC has worked with the New York Department of Transportation to widen Route 300 along a 2.2-mile section in Newburgh, Orange County, adjacent to both the New York State Thruway and Interstate 84.

Clearly, J. Fletcher Creamer & Son, Inc., has a stake in New Jersey, but what about the future? "New Jersey has had a boom in construction, heavy in both highway and residential," says Creamer. "There have been a tremendous amount of office buildings constructed in the last six years or so. We're experiencing the biggest growth that we've ever had in this state to my knowledge, and I have been here a

long time. I believe that will continue through the turn of the decade." In commenting on the strides the state has made with the infrastructure, Creamer says, "many things are happening. The turnpike will be widened from exit 9N to six lanes. The Garden State Parkway is improving its service areas, providing more parking and better ingress and egress, and the DOT's ongoing program has been replacing many bridges in need of attention. So we are doing much to update."

Becton Dickinson and Company—Becton Dickinson is another example of a company whose roots go deep into New Jersey soil. The *Fortune* 500 company is among the health care giants. It was begun by Maxwell Wilbur Becton and Farleigh Stanton Dickinson in 1897 in New York City. From imported fever thermometers, their line quickly grew to include hypodermic units and other supplies. The company was incorporated in New Jersey in 1907, the same year it was moved permanently to a new factory in East Rutherford, New Jersey, in order to ensure consistent product quality and supply. This became one of the first United States plants to manufacture hypodermic units.

Over the years, Becton Dickinson has been a leader in its field. During World War II, it produced all-glass syringes which supplanted and improved upon the metal units being used. The company also

developed the "all cotton elastic" bandage, better known as the ACE ®
brand bandage. Its pool of brilliant inventors have kept Becton
Dickinson in the avant garde of medical products research, among
them Andrew W. Fleischer and Dr. Oscar Schwidetzky. Fleischer modernized the stethoscope and invented the first accurate instrument to
measure blood pressure. Schwidetzky developed special hypodermic
needles and other medical devices.

During World War II, Becton Dickinson was awarded an Army-
Navy "E" for Excellence, in recognition of its vital role in providing
quality medical equipment for the war effort. During the 1940s,
Becton Dickinson established Fairleigh Dickinson Junior College,
which later became Fairleigh Dickinson University.

After the war, the company continued its innovation by developing such new products as the glass syringe with interchangeable parts
and a device to collect blood in sterile tubes sold under the
VACUTAINER ® brand.

From the 1950s to the 1980s, Becton Dickinson went through a
period in which it acquired many leading firms, and by 1961 the company introduced a product that would alter its future forever. With the
development of the disposable hypodermic syringe, the company

Raymond V. Gilmartin (right),
president of the Becton Dickinson
Corporation, stands beside the
Chairman and Chief Executive Offi-
cer, Wesley J. Howe. Photo by Mark
Ferri Photography. Art directed by
William Snyder Design, NYC. Cour-
tesy, Becton Dickinson Corporation

CPC INTERNATIONAL INC.

James R. Eizner, chairman, president, and chief executive officer of CPC International Inc.

Headquartered on Sylvan Avenue in Englewood Cliffs, CPC International Inc. has annual sales of close to $5 billion, ranking the company among *Fortune*'s 100 largest industrial corporations in the United States and fourth largest in New Jersey.

In the food industry, CPC International is one of the dozen largest food-processing companies in the United States. Of them all, CPC is by far the most international, deriving more than half of its revenues from operations abroad. Worldwide, CPC has 102 manufacturing plants in 47 countries.

CONSUMER FOODS

CPC's consumer foods business accounts for more than 80 percent of the company's total sales. Its major brands generally occupy the number-one or a strong number-two market position in their respective categories in the countries where they are produced.

Best Foods, CPC's North American consumer foods business, is also head-

quartered at Englewood Cliffs. Best Foods has annual sales approaching $2 billion, and produces and markets many long-established household favorites and a number of new products. Among Best Foods' well-known products are: Hellmann's™ and Best Foods™ mayonnaise, Mazola™ corn oil and margarine, Skippy™ peanut butter, Thomas'™ English muffins, Arnold™ breads and rolls, Old London™ and Devonsheer™ melba toast, Mueller's™ pasta products, Karo™ and Golden Griddle™ syrups, Knorr™ soups and sauces, Argo™ and Niagara™ laundry starches, and Rit™ tints and dyes.

In New Jersey, Best Foods has, in addition to its Englewood Cliffs headquarters, a major research laboratory in Union, and plants in three locations, including Bayonne—where a large new plant produces Hellmann's™ mayonnaise and Mazola™ corn oil; Jersey City—largest of the Mueller's pasta plants; and Totowa—the Best Foods Baking Group's largest plant for Thomas'™ English muffins.

Abroad, CPC International has consumer foods divisions in Europe, Latin America, and Asia. Worldwide CPC produces Knorr dehydrated soups, sauces, and bouillons for major markets in 37 countries; mayonnaise for 28 countries, chiefly under the Hellmann's™ and Best Foods™ brand names; Mazola™ corn oil for 29 countries; Maizena™, Duryea™, Brown & Polson™, and other packaged corn starches for 36 countries; Alsa™, Lady's Choice™, and Majala™ desserts; Adler™ and Bavaria™ cheese products; Hirz™ dairy products; and many others.

CORN REFINING

CPC's corn-refining operations, which account for roughly 18 percent of its worldwide sales, serve a broad industrial customer base of more than 60 different industries, providing them with corn sweeteners—dextrose, high-fructose corn syrup, and glucose syrup; regular, modified, and special corn starches; corn gluten feed and meal; and corn oil.

CPC's worldwide Corn Refining Division has operations in the United

The headquarters of CPC International Inc. and its Best Foods Division, located in Englewood Cliffs. The flags in the foreground represent a few of the 47 countries in which CPC International Inc. operates.

States, Canada, and eight countries of Latin America. It also has a cooperative management group that coordinates and manages licensing agreements and minority interests in other countries.

CPC's North American corn-refining business has three plants in the United States and three in Canada. Five of these are totally new, having been constructed since 1980, and the other has been thoroughly modernized and expanded.

HISTORY

CPC International's history began in 1906, when several corn-refining companies combined to form the Corn Products Refining Company. For 52 years it continued to operate primarily as a corn-refining company, but also making such branded food products as Mazola™ corn oil, Karo™ corn syrup, and Argo™ and Kingsford's™ corn starch.

In 1958 significant strategic changes were made when Corn Products Refining Company merged with Best Foods, Inc., which had a number of important consumer products, including Hellmann's™ and Best Foods™ mayonnaise, Skippy™ peanut butter, and Rit™ dyes. The corporate name was changed then to Corn Products Company. At the same time the organization acquired C.H. Knorr A.G. in Germany and Switzerland, adding an important line of Knorr soups then produced in four countries.

In 1969 the company name was changed to CPC International Inc. The new name signaled both the significant growth of the firm's consumer foods business and the strong expansion of its international business.

As the company grew its sales crossed the billion-dollar mark in 1966 and reached $4 billion in 1980. CPC's growth came chiefly from additions to its product lines, geographic expansion, and acquisitions. For example, Knorr™ soups were marketed in only four countries when that business was acquired by CPC in 1958; today Knorr™ products are made in 36 countries and include more than 600 varieties of soups, sauces, bouillon cubes, seasonings, base mixes, snack meals, and other products. Similarly, CPC's mayonnaise business began in the United States, and today the firm produces mayonnaise in 20 countries for markets in 28 countries.

The organization's early history includes legends of many innovative entrepreneurs, such as Samuel Bath Thomas, a baker who established a shop in New York in 1880 and began a business that developed Thomas'™ English muffins. Thomas'™ products are made in plants in New Jersey, Maryland, California, and Illinois, and are now available to more than 70 percent of U.S. households. Also, there was Richard Hellmann, whose New York delicatessen first sold Hellmann's™ mayonnaise in 1912. His resealable glass jars were ornamented with a blue ribbon, just as jars of Hellmann's™ mayonnaise are today.

This interior elevation is of the Becton Dickinson headquarters in Franklin Lakes. Photo by Steve Rosenthal. Courtesy, Becton Dickinson Corporation

entered a new era. Differing significantly from the traditional reusable needle, the disposable syringe required huge investments in new process and sterilization equipment, packaging, and distribution. To raise capital, Becton Dickinson became a publicly-held corporation. In 1963, the company was listed on the New York Stock Exchange. Within a year, the new disposable syringe took over one-third of the reusables market. By 1987, reusables accounted for only a fraction of one percent of the market.

Since then, Becton Dickinson has continued to uphold its reputation as a giant in the field of modern medical technology and products. As a leading diversified transnational health care corporation, it has sales of more than $1.5 billion and more than 19,000 employees at 74 locations in 20 countries.

In 1986, Becton Dickinson moved into its new headquarters in Franklin Lakes. The distinctive new brick and stone building, which has garnered accolades for its architectural uniqueness, will house all company operations by the turn of the decade.

Union Camp Corporation—Ranking among the top 200 industrial companies in America, Union Camp Corporation's principal products are paper, packaging, chemicals, and building materials. The company has manufacturing facilities in 22 states, international operations in 23 countries, and 1.75 million acres of woodlands in the Southeast. The packaging group integrates the company's Kraft Paper and Board Division and all of Union Camp's packaging divisions, whose raw materials are mostly Kraft paper and board. The Fine Paper Division is the second major operating group. It produces white paper and coated and uncoated board. The chemical group consists of the Chemical Products Division and the Bush Boake Allen Division. The latter combines Bush Boake Allen—a worldwide producer of flavors and fragrances acquired by Union Camp in 1982—and the company's previously designated Terpene Chemicals Division. The Building

Products Division supplies the home improvement and construction markets, and also serves industrial manufacturers with specialized wood products.

Union Camp Corporation is the result of the 1956 merger of Union Bag and Paper Corporation and Camp Manufacturing Company. Union Bag and Paper Corporation was the outgrowth of several other companies, the oldest being Union Paper Bag Machine Company. This company was formed in 1861 by Francis Wolle, a Moravian minister and teacher, who, in 1852, invented the first paper-bag machine. Camp Manufacturing Company dates back to 1887 with the purchase of a lumber manufacturing operation that predated the Civil War.

Union Camp Corporation moved its headquarters from the Woolworth building in New York City to Wayne, New Jersey, in 1969. In an interview, Tom Hunter, director of public relations for Union Camp, discussed New Jersey's advantages:

We had been in Manhattan for decades, but New Jersey's amenities were attractive. We started with 50 potential sites mostly within the greater metropolitan area and also in the south and Middle Atlantic states where we have a lot of operations. We chose Wayne as the most convenient spot because we thought there was still a value in being close to New York.

Union Camp's assistant director of public relations, Timothy McKenna, pointed out that, "…We don't need to be as close given the electronic communications but it helps to be near the excellent legal, financial, and creative resources that are available to a big company like Union Camp."

Hunter pointed out that the general amenities available in a suburban setting were important: "Better surroundings, a highly educated and abundant labor pool, the lower cost of doing business, and the more pleasant commute were all major factors," he said. "When we were in New York, many of our people were compelled to wrestle with the Long Island Railroad and heading up into Westchester, Connecticut or New Jersey." He went on to stress that now Union Camp can house 425 employees at the headquarters location. "Having everyone from the corporate family together is important. We wouldn't be able to get a facility like this one in a place like Manhattan." Union Camp's striking office facility in Wayne is positioned at the edge of a beautiful reflecting pond. A feeling of family and camaraderie is exhibited at the company. "People here know each other on a first name basis; it's a family

ERNST & WHINNEY

Receptionist Lee Alberti greets clients in the Ernst & Whinney lobby.

One of the world's preeminent professional service organizations and a prominent member of the Big Eight public accounting firms, Ernst & Whinney has served Bergen and Passaic counties and other rapidly growing areas of North Jersey from offices in Hackensack since 1974.

From its headquarters in Continental Plaza, the worldwide resources of the firm's network of some 350 offices can be channeled to serve the needs of large and small clients in the region. Conversely, the office serves as a conduit to the expertise of Ernst & Whinney's more than 20,000 professionals for clients expanding into new areas of the United States and around the world.

The Hackensack office, one of seven in the New York-New Jersey metropolitan area, has 61 professionals who provide clients with the full extent of Ernst & Whinney's audit and accounting, tax, and consulting services, as well as personal financial and estate planning. The firm brings a diversity of services and capabilities that can be applied to the individual needs of clients, be they small, highly entrepreneurial privately held businesses, or huge *Fortune* 500 corporate units.

It is the rise of high-technology, entrepreneurial, relatively new companies that has fueled the tremendous expansion of the North Jersey market and powered the growth of Ernst &

Whinney's operations in the region. In recognition of this particular market's ongoing needs, the firm has instituted a group specializing in Privately Owned Emerging Businesses (POEB). While these clients may not require the full range of Ernst & Whinney's services immediately, they have demonstrated

Five of the partners in the Hackensack office of Ernst & Whinney are (seated from left) Louis Feuerstein, John Kilkeary, and Michael Wilk, and (standing from left) Jan Chason and Richard Scherger.

real potential for growth and can be considered candidates for development and the full services of the firm in the future.

In addition to the growth of new businesses in the northern New Jersey market during the latter half of the 1980s, another dramatic impact on the prosperity of the region has been the migration of major corporate headquarters out of New York City to communities throughout Bergen, Passaic, and, more recently, Morris counties. These have been, generally, more mature businesses requiring the fuller service offerings of the firm.

The combination of locally grown, newer, and highly entrepreneurial companies with the corporate migrants from New York to the region resulted in a truly dynamic enconomic entity encompassing the entire northern New Jersey region. Bergen and Passaic counties alone constitute the 27th-largest market in the top 50 rated by Ernst & Whinney throughout the United States.

As a full professional service organization, Ernst & Whinney furnishes the guidance and capabilities required by all strata of businesses, governmental agencies, and institutions to maintain

Partners Jan Chason, John Kilkeary, and Louis Feuerstein discuss potential service opportunity.

and enhance their abilities to fulfill their individual missions and exploit opportunities in an expanding marketplace.

The worth of Ernst & Whinney, however, is the value the firm's services add to its clients' ultimate results. And this can only be achieved by Ernst & Whinney's commitment to improving the scope and depth of its own services.

Certainly key to this objective is the recruitment, development, and retention—through exceptional career opportunities—of highly talented personnel who bring a high degree of creativity to their jobs.

With the ever-broadening applications of high technology to business, information system consulting has become a key developing service area— one that demands imaginative, creative individuals to provide the consulting services and software development that are concomitants of modern business procedures.

Another important aspect of the firm's services may not be obvious, or even listed in its marketing literature. Rather, it is the value of the firm's long experience in matters involving mergers, acquisitions, geographic expansion, and the benefit of being able to refer cli-

ents to effective legal, financial relations, and other counselors to assist with a company's strategic programs.

These skills and capabilities are brought together for the client's specific needs. Ernst & Whinney focuses its services through the integration of audit, tax planning, and management consultation programs into a comprehensive partnership between the firm and each client.

The broad client base of Ernst &

Whinney's Hackensack office testifies to the success of the firm's intense client orientation. Major local and national manufacturing, computer software, cable television, leasing, distribution, publishing, and construction companies are numbered among the office's client list.

As northern New Jersey enters the final decade of the twentieth century, Ernst & Whinney is poised to broaden its base in the region further through expansion of its existing operations and the possible acquisition of quality local and regional accounting firms.

"This is clearly a premium market for our firm in the years ahead," says managing partner John Kilkeary. "It is a region with tremendous community spirit and involvement on the part of companies located here. Our task is to build on the level of service Ernst & Whinney has given in the past to meet the new and broad demands of the future."

Ernst & Whinney's nationwide and globe-circling expertise and capabilities are teamed with the local responsiveness of its Hackensack office to provide a key building block for the future success of locally based companies.

Partners Michael Wilk (left) and Richard Scherger consulting on tax strategies for their clients.

BECTON DICKINSON AND COMPANY

Becton Dickinson and Company has been a major New Jersey corporation for most of its history, providing high-quality, cost-effective health care products for doctors, hospitals, medical laboratories, and consumers throughout the world. Organized in New York City in 1897 by salesmen Maxwell Wilbur Becton and Fairleigh Stanton Dickinson, the company was incorporated in New Jersey in 1907, the same year it moved permanently to a new factory in East Rutherford, in order to ensure consistent product quality and supply. The plant was one of the first U.S. facilities to manufacture hypodermic needles.

In World War I the firm produced all-glass syringes, a significant improvement over the metal units of the day. Becton Dickinson also developed the ACE® brand bandage, which stood for "all cotton elastic."

During World War II the company was awarded an Army-Navy "E" for Excellence in recognition of its role in providing quality medical equipment for the armed forces. During that time Dickinson and Becton played key roles in establishing what is now Fairleigh Dickinson University.

With the help of products ranging from culture dishes to automated instruments, physicians can often make a complete diagnosis in just a few minutes. The qualities of Becton Dickinson's diagnostic products make for fast, cost-effective medicine.

Following the war a stream of innovative diagnostic and medical products came from Becton Dickinson, including a glass syringe with interchangeable parts and a device that collects blood in sterile tubes for laboratory analysis.

After the founders died, the company continued to broaden its product lines and to build new production capacity while expanding its international business in Canada, South America, and Europe.

Over the years Becton Dickinson's commitment to New Jersey has continued to grow. There are now more than 2,400 employees in manufacturing and administrative office locations in the state. The Becton Dickinson division, which produces syringes and needles, is located in East Rutherford. The Ivers-Lee division, a major supplier of single-unit flexible packaging and packaging machinery, has been in West Caldwell since 1969. Becton Dickinson Phar-

maceutical Systems, a leading producer of syringes to pharmaceutical companies for prefilling with therapeutic drugs, also is located at the West Caldwell site. Becton Dickinson Infusion Systems, a principal supplier of advanced syringe pumps, is in Lincoln Park.

In 1986 Becton Dickinson moved into an architecturally acclaimed new brick and stone headquarters on a 130-acre wooded site in Franklin Lakes. When a second building is completed shortly after the turn of the decade, all of the company's administrative operations in New Jersey, including those now in East Rutherford, will be consolidated in Franklin Lakes.

The firm maintains a high regard for East Rutherford and intends to ensure that its property continues to be a valuable commercial asset for the benefit of the community after the East Rutherford administrative offices are transferred to Franklin Lakes. In anticipation

The concern for safety, ease of use, and efficiency is reflected in the qualities of the Becton Dickinson medical products used in hospitals, physicians' offices, and homes to deliver health care around the world.

The BACTEC®-brand blood culture system screens blood samples for the presence of microorganisms.

of this move, Becton Dickinson has been developing plans for alternative uses of the site. The first step in the ultimate conversion of the property was taken early in 1988, when the company sold 13.2 acres to the Federal Reserve Bank of New York for a future operations center.

Today Becton Dickinson is a leading diversified, transnational health care corporation with sales of more than $1.5 billion and more than 19,000 employees at 74 locations in 20 countries.

The firm is organized in three sectors. The Diagnostic Sector develops and manufactures sensitive diagnostic products and systems that quickly and precisely provide information physicians need to diagnose disease and prescribe therapy. Its products are found in all stages of the diagnostic process, from specimen collection to computerized instrumentation that helps physicians and laboratory specialists analyze the specimens.

The Medical Sector produces dispos-

able syringes and needles, thermometers, protective gloves, and elastic products, as well as improved ways to introduce therapeutic drugs into the body, systems to monitor the critically ill, a full line of products for operating rooms, and products that enable people to care for themselves in the home.

The International Sector makes and markets high-quality Becton Dickinson products that are adapted to the needs of each country where Becton Dickinson

does business. Modern plants and distribution and training centers in all major world areas testify to the company's long-standing commitment to the international health field.

Becton Dickinson's future success will continue to derive from its technological leadership in products, manufacturing processes, and services. The firm's worldwide research and development is conducted by more than 700 technical personnel, more than 300 of whom hold advanced degrees in a broad array of scientific disciplines. Researchers in the organization's divisions work to improve products and processes. Longer-term fundamental research is carried on at two major facilities: the Corporate Research Center in Raleigh, North Carolina, and the Becton Dickinson Monoclonal Center in Mountain View, California.

Becton Dickinson and Company's attention to quality and service is basic to all corporate operations and to its people, who are dedicated to providing a high level of service to its worldwide customers as the firm looks forward to its second century in New Jersey.

Technological innovation means finding new ways to improve the performance of traditional products and improve manufacturing processes. Becton Dickinson basic and applied research laboratories utilize a wide range of scientific and technical disciplines, aided by computer-assisted design programs, to carry out projects ranging from modifying polymers to studying the structure and function of human cells.

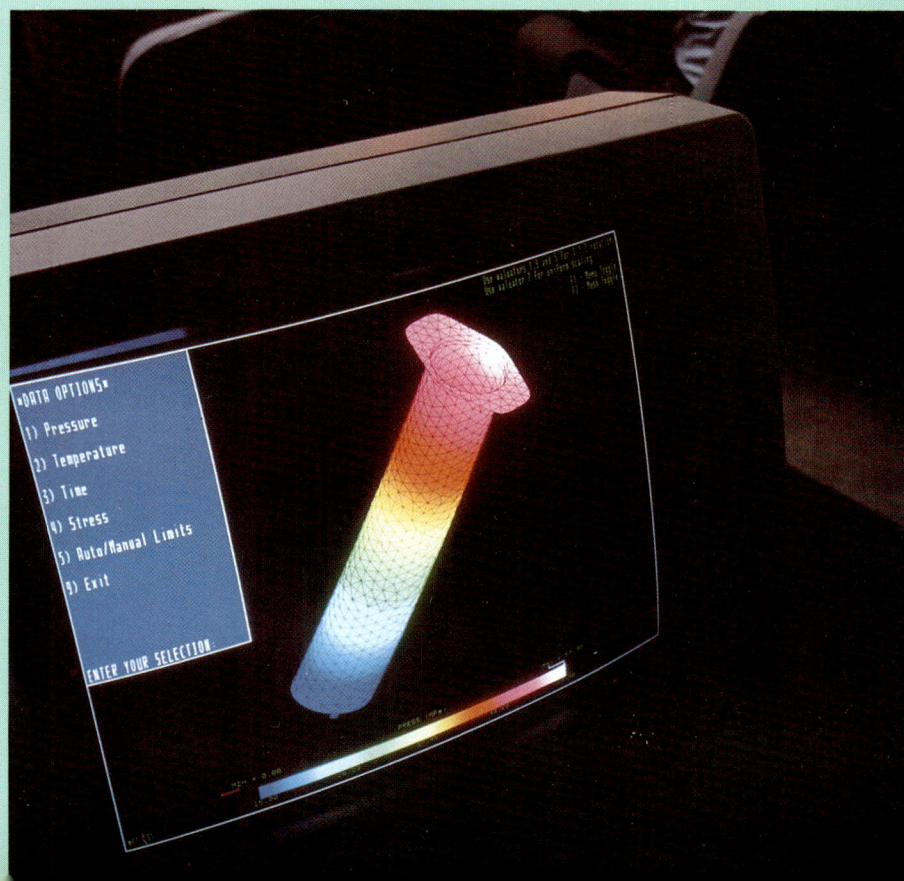

feeling," Hunter noted, "and I'm sure part of that is attributable to our work environment. I think being in a suburban community fosters that. An office building in New York is not in a community."

Interaction with the community is inevitable in a suburban setting. Corporations are literally "residents" of the community in which they do business, and Union Camp is an active member of its community. "We interact quite closely with the school district. Our cafeteria is frequently used by community groups for meetings. In fact," Hunter said,

we have even had secretaries who have had wedding receptions here. Our people also participate on the boards of various institutions in the community. We also have a community relations person who is deeply involved in local interaction. We have people here who get out and contribute a lot of time to their towns. I think ordinary people here care.

In summing up on the topic of northern New Jersey's assets, McKenna says, "I think New Jersey's strongest suit is its work force, which is highly educated on every level—even clerical workers have gone to college for two years, or secretarial school. There are plenty of computer programmers, many accountants, and also creative people." He continued, "I know many other areas of the country, so I can say with some authority that they don't have the quality work force New Jersey has, and it hurts them."

Commenting on the quality of life in New Jersey, McKenna says, "the state is much more than the 'turnpike' or Hudson County anymore. There is a thriving economy and a good life here. I think most people outside New Jersey are aware of that now." Hunter added, "culturally, you get things here that you wouldn't in any other state. You can see almost anything here because New Jersey is a major stop for most tours and New York City is 'next door.' I think children in this area are lucky. Their lives and minds can't help being enhanced intellectually."

Union Camp's northern New Jersey facilities include a folding carton plant in Clifton; a printing plant in Englewood; a plant in Moonachie, which was acquired this year; headquarters for the London-based subsidiary Bush Boake Allen, in Bergen and Montvale; and Bush Boake Allen's plant in Norwood. The company's research and development facilities in Princeton are in the middle part of the state,

and several other institutions are in Trenton and the southern part of New Jersey.

CPC International—This firm, headquartered in Englewood Cliffs, ranks among *Fortune*'s 100 largest industrial corporations in the United States. Its annual sales are close to $5 billion. CPC is one of the 12-largest food processing companies in the country and the most international of them all—more than one-half of its revenues are derived from operations abroad. Worldwide, CPC has 105 manufacturing plants in 47 countries.

Best Foods, CPC's North American consumer foods business, is also headquartered in Englewood Cliffs. Its annual sales approach $2 billion. The company produces and markets such well-known products as Hellman's and Best Foods mayonnaise, Mazola corn oil and margarine, Skippy peanut butter, Thomas' English muffins, Arnold breads and rolls, Old London and Devonsheer melba toast, Mueller's pasta products, Karo and Golden Griddle syrups, Knorr soups and sauces, Argo and Niagara laundry starches, and Rit tints and dyes.

In addition to its Englewood Cliffs headquarters, Best Foods has several other northern New Jersey locations, including a major research laboratory in Union; a large new plant in Bayonne, which produces Hellman's mayonnaise and Mazola oil; the largest of the Mueller's pasta plants, in Jersey City; and the home base for Best Foods Baking Group and a plant for Thomas' English muffins in Totowa.

CPC's consumer-foods business accounts for more than 80 percent of the company's total sales. Its corn refining operations, which accounts for approximately 18 percent of its worldwide sales, serve a broad industrial customer base of more than 60 different producers.

OVERSEAS CORPORATIONS AT HOME IN THE NORTH
Foreign corporations have always found northern New Jersey a great place to locate. The state houses 1,000 companies from 48 countries, and many of those firms can be found in the north. Employees of these companies also find northern New Jersey an agreeable place to live. According to Gordon Bishop in his excellent account of the state,

These well-known food products represent the Best Foods division of C.P.C. International Inc. Courtesy, C.P.C. International Inc.

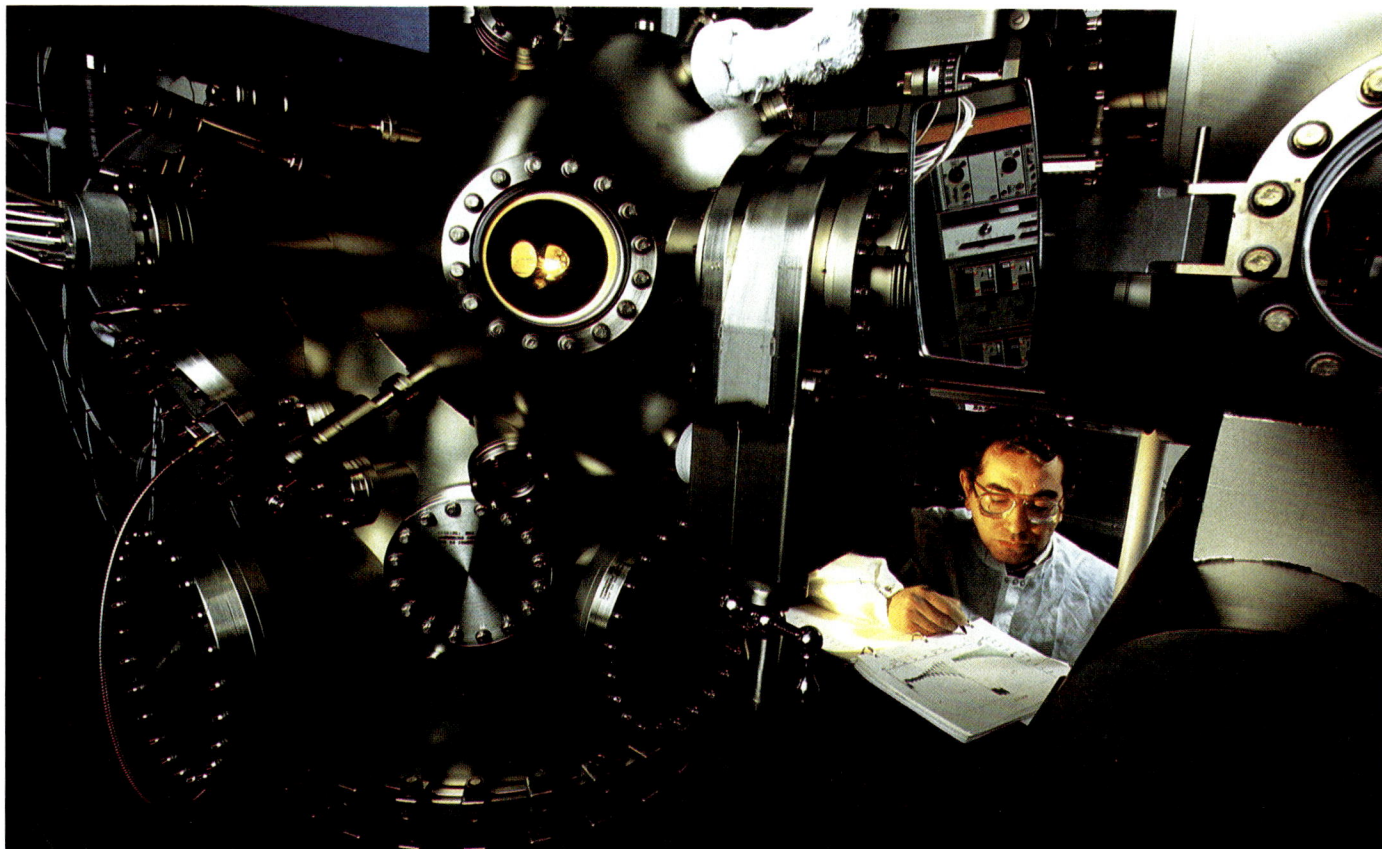

Siemen's Laboratory conducts research in Epotexy. Photo by Bob Krist

Gems of New Jersey, many are attracted by New Jersey's geological diversity and its fiscally exceptional Triple-A bond rating. "More than 5,000 Japanese nationals have settled in the Bergen-Hudson-Essex region," Bishop writes, "representing firms such as Panasonic, Toyota, Datsun, Sharp, Minolta, Fuji, Yashica, and an impressive array of high-tech companies."

The BOC Group, Inc.—One company that is happy to have a northern New Jersey address has over 100 years of leadership within its market. The British company, The BOC Group, Inc., has as its largest area of activity—and the one it is most traditionally associated with—the production and supply of all major industrial and specialty gases and related products. From humble beginnings, the industry has expanded into a worldwide business with sales of over £7 billion.

The BOC Group is among the few international companies that dominate the market. Since BOC began producing oxygen, the list of gases that it produces and markets has burgeoned to include half-a-dozen commonly used gases and thousands of special gases and mixtures. Today, The BOC Group is one of the largest within its market in the world, employing 40,000 across the globe. At the time of its 100th anniversary in 1986, the BOC Group had sales of industrial gases and cryogenic materials at around £1 billion.

Health care now constitutes the company's second-largest area of activity. This branch had its origins in the supply of anesthetic gases, pharmaceuticals, and related equipment. Now health care products extend into many other areas of patient care.

BOC's carbon-based products are primarily marketed in the steel industry, which is also a major consumer of industrial gases. Of the group's carbon-based businesses, the largest is Airco Carbon, whose major product line is graphite electrodes.

As a result of the company's position as a leading supplier of welding gases, it has pursued interests in welding equipment and materials. The BOC also has significant interests in vacuum engineering, carbide products, and educational and food services.

In addition to the ever-increasing stable of internationally renowned companies, like The BOC Group, that have chosen northern New Jersey locales, many foreign automakers have selected Bergen County as their North American headquarters. With its proximity to Port Newark, the area is a prime location.

Throughout the 1980s, New Jersey has been a leader in automobile importing. Over 300,000 foreign cars a year have been received through Port Newark. With its concentration of foreign car makers, Bergen County stands out as a mecca for these international giants. It has prompted many to call New Jersey the foreign-car capital of the United States.

BMW of North America—One foreign automaker that has had a pleasant tenure in northern New Jersey dates its history in the area back to 1975. BMW of North America in Montvale was established by the parent company located in Munich, Germany. After purchasing its importer distributorship from Max Hoffman, who had been importing BMW, Mercedes, Saab, Porsche, Volvo, and many other foreign cars for many years, the company was on its way in the United States.

Hoffman, the original importer of all European makes, set down roots in Bergen County—his home. It seemed only logical for numer-

THE BOC GROUP

The BOC Group Technology Center in Murray Hill, New Jersey.

Northern New Jersey has advanced dramatically in its economic and social development since the end of World War II. The northernmost counties of the state today boast the headquarters and major regional offices of many of the nation's—and indeed the entire world's—most prestigious business and financial organizations, located within vibrant communities well known for their residential, recreational, and cultural amenities.

It is not surprising then to find the U.S. headquarters of The BOC Group there, relocating in early 1989 to Mur-

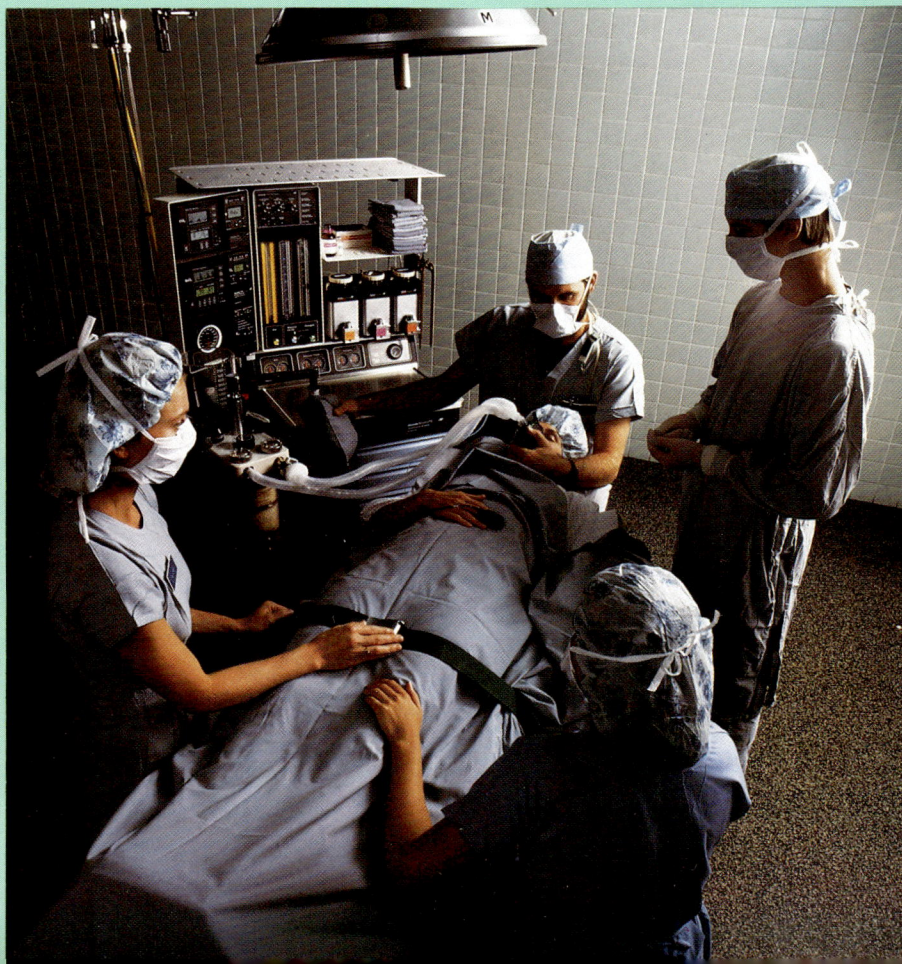

The Modulus II anesthesia machine is an example of the medical equipment for critical care that is produced by the BOC Health Care division of Ohmeda.

ray Hill from Montvale, where it had been domiciled since 1972.

The BOC Group is the quintessential U.S.-British hybrid business organization, led by an American chief executive officer, Richard V. Giordano, but with its origins decidedly in the United Kingdom, where its corporate headquarters is centered, with operational and research facilities located in the United States, Europe, and elsewhere.

The BOC Group is not diversified only from a geographic standpoint, but from the broad base of its enterprises as well. The company has assembled an international and world-competitive portfolio of businesses.

Of its three primary fields of business—gases, health care, and special products and services—atmospheric gas extraction (oxygen, nitrogen, argon, etc.) and production of synthetic gases are the dominant activities of the group's companies. BOC is one of only two corporations that compete on a truly international basis in the field of industrial gases. Its gases are found in all kinds of products and processes. From light bulbs to margarine, frozen food to microchips, BOC plays an invisible yet essential part in other companies' manufacturing processes.

BOC's gas businesses extend quite literally around the world, with sizable emplacements in North and South America, in Europe, and in Africa, while fully 40 percent of its activity is now in the high-growth Asia/Pacific region.

While BOC companies are leading producers and marketers of gases and related technologies, the group is heavily invested in research and development centered at the Group Technical Center in Murray Hill. As a leading producer and distributor of medical oxygen and a manufacturer of anaesthesia equipment, The BOC Group's expansion into the health care field is a closely aligned extension of its traditional business operations.

Expanding upon the medical oxygen base, the group's health care businesses specialize in four distinct areas:

BOC businesses stay at the forefront of their respective fields through constant technological research and innovation.

pharmaceuticals, medical equipment and systems for critical care and medical engineering, intravenous disposables, and home health care. The group also has other interests in the health care field, from making and selling medical gases in many countries, to operating an Australian medical distribution service.

Health care operations make up the group's fastest-growing business sector, providing more than 25 percent of BOC profits. The BOC Group plans to expand in the critical and acute care sector by developing systems, products, and services through the growth of its existing businesses; acquisitions; co-marketing; moving into complementary areas; and diversifying into new ones.

The group's Special Products and Services businesses derived from BOC's principal activities in gases and health care, both fertile grounds for the development of new technologies and the refinement of management skills that are applicable to other businesses.

This small but dynamic element within the BOC lineup of companies maintains its forefront position in the distribution and high-vacuum technology fields through constant research and innovation.

The group's distribution services are a

direct outgrowth of its gases operations. Distribution is a principal key to success in the gases industry, and BOC companies have achieved a formidable reputation in the field based on their superior performances. BOC Distribution Services has grown out of sophisticated storage and distribution expertise honed through the extensive use of computers and other technologies.

With regard to advanced technologies, BOC's entry into high-vacuum

equipment and processes dovetails with the rapid adoption of advanced vacuum technology by industry in the manufacture of all kinds of products, from foods to pharmaceuticals. These developments have presented technical challenges to BOC's two subsidiaries in the high-vacuum technology field. They have met these challenges with technical and consequent commercial success.

With its widespread, diversified lines of businesses and operations in 50 countries employing some 37,500 people, including representatives in every state of the union, The BOC Group requires relatively easy access for management personnel as well as logistical support. New Jersey, particularly its northern counties, provides the centralized link needed within this far-flung organization.

Thus, the group has 11 offices or facilities in northern New Jersey—including the U.S. corporate offices, The BOC Group Technical Center, and headquarters for two divisions.

In summing up the region's attraction to The BOC Group, chairman and chief executive Richard V. Giordano said: "The area offers us a high quality of life both from the standpoint of business operations and for our employees at home. Northern New Jersey has lived up to our expectations."

BOC's gases are used to make virtually everything that supports modern life—from food to pharmaceuticals, from metals to microchips.

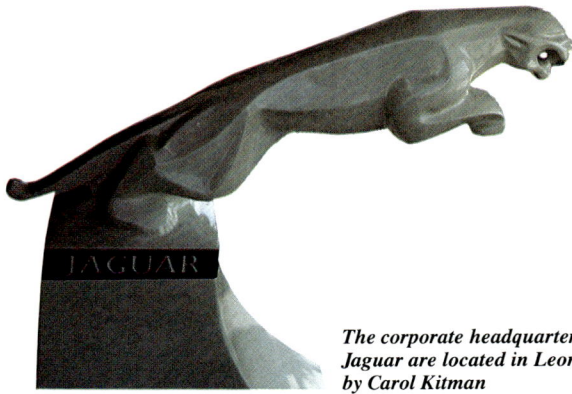

The corporate headquarters of U.S. Jaguar are located in Leonia. Photo by Carol Kitman

ous reasons for the foreign company to follow suit when establishing a home base.

Richard S. Brooks, Jr., corporate communications manager for BMW, outlined some of the reasons for the company's long residence in New Jersey: "There are numerous reasons why New Jersey is ideal for us," he said, but the two topping the list are space and quality

of life issues. As a headquarters facility, we need space. Up to two years ago we had a parts department in this building, a warehouse for the entire East Coast (west of the Mississippi), where we kept parts for over 200 dealers. There were products that ranged from auto and motorcycle parts to lifestyle accessories: jackets, shirts and other clothing.

Easy access to both New York City and Philadelphia as well as the quality of life available makes northern New Jersey a highly desired location for both private and corporate residence. Photo by Michael Spozarsky

Brooks explained the unique needs that determine space requirements for auto companies:

…we do a lot of our training here because it is where the information resides. We have all our computer equipment here, which requires a large amount of space also. We computer monitor all our dealers on the mainframe. We don't have any back office; it's all done here at the headquarters facility. This ends up a very central place. You can't get that in New York City, where space is at a premium; yet, we are close enough to the city for things like marketing.

In regard to quality of life issues, Brooks asserts that they are paramount:

Last year, the president of BMW, who lives in Upper Saddle River and likes it very much, maintained that one of the main reasons for staying is that the quality of life here is so high. He feels that the area is exciting and dynamic. There is so much going on and so much that will happen in the future that a comparable value would be hard, if not impossible, to find.

Brooks points out that "the quality of life is important to our employees, and they are happy here. It's nice to go out to a park on your lunch hour. The area is special because it has a lot of advantages over other places. Living and working here means that New York City, the mountains, and lakes are all only a half hour away."

These two reasons were major factors in making the decision to stay on in Bergen County. This year, BMW will consolidate its three buildings and move its operations across the Garden State Parkway to Chestnut Ridge Road, also in Montvale. The company will be in a "neighborhood" of other international giants, among them The BOC Group, Inc. "We've decided to stay here because the area has served us well," Brooks said. "We want to stay in the region in which we grew best. When BMW of North America started, we sold 14,000 cars a year; now we sell almost 100,000. We expanded very quickly." With the new facility, BMW will own its own building. "We rent this one; now we can house everyone under one corporate roof. That's very important to us," Brooks said.

Commenting on the ability of foreign personnel to acclimate to northern New Jersey's lifestyle, Brooks said,

The president is German [as are] three or four vice presidents in various departments and some technical engineers. They are all quite happy living in New Jersey. Their kids go to school here, and our CEO's wife is very active in community affairs in the identical way any CEO's wife would be. The style of life here is so agreeable, acclimation posed no problem.

In building bridges with the community, BMW, like most of the major corporations, makes a point of getting involved. "The March of

Mr. Bjorn Ahlstrom is president and chief executive officer of the Volvo North America Corporation. Courtesy, Volvo North America Corporation

Dimes is one of the biggest things here," Brooks said. "The Chestnut Ridge Harvest Ball is a very successful annual event. There is an upscale auction of a Mercedes and a BMW." Brooks underscores the involvement with the local school districts, "which benefits the schools and the company. We also recognize that the community is concerned about the strain corporations put on local police and ambulance squads. We are now planning to work out a way companies can donate money or employee volunteer time. Let's face it, we need each other and we both realize it."

In a final comment on the northern New Jersey area in which BMW resides, Brooks talked about its success: "I think they just happened to have all the things in place that people wanted. They don't have to go out of their way to lure people to New Jersey as many other states do by offering incentives; they don't have to. New Jersey has what it takes. People come here on their own because they want to."

Mercedes-Benz of North America—As BMW's neighbor in Montvale, Mercedes-Benz moved to its location in 1972. The impressive headquarters building has 145,000 square feet of floor space, and it is situated on a 23-acre site. The company employs 1,700 people throughout the United States. More than 600 of these are housed in the three Montvale facilities. In addition to its headquarters facility, there is a 123,542-square-foot building situated on 10 acres, which is utilized as a computer center, office, and service training center. The New York Zone headquarters, warranty inspection, and technical test and service center occupies a 42,300-square-foot building on four acres.

Mercedes-Benz of North America, established in 1965, is a wholly owned subsidiary of the world's oldest automotive manufacturer: Daimler-Benz AG of West Germany. In 1987, Mercedes-Benz of North America sold 89,918 passenger cars through its 423 United States dealers.

Volvo North America Corporation—Volvo established its first North American company in April 1956. Since that time it has grown from a small importer of automobiles into a diversified international group of companies operating primarily in the areas of transportation equipment, energy, and food.

North America is Volvo's largest single market, accounting for approximately one-third of the total sales. The parent company, AB Volvo, is Scandinavia's largest industrial enterprise, generating more than $11 billion in annual sales and employing more than 70,000 people worldwide. Volvo employs 800 people in New Jersey. The company is located in Rockleigh in Bergen County.

Other Bergen County foreign automakers include Rolls Royce and Peugeot in Lyndhurst; Jaguar and Rover in Leonia; Alfa-Romeo and Renault in Englewood Cliffs; and Citroen in Englewood.

A HEALTHY JOB MARKET The influx of corporations to northern New Jersey has been a boon to the area and the state as a whole. Northern New Jersey's successful entrance into a service sector economy has set trends for the rest of the country. In turn, the growth and expansion of the region has created abundant job opportunities. These employment needs have been met by the area's highly educated and skilled work force. Economists called 1987 the best year for New Jersey in two decades, with employment surpassing the 1986 all-time high. Labor market conditions were the best since the full-employment economy during the Vietnam War. According to E. James Ferland, Public Service Electric and Gas board chairman and chief executive officer, in an article in the *Star Ledger* Outlook edition: "The service side of New Jersey's economy has become the state's strong suit in recent years; since the end of the last recession in 1982, employment in services—wholesale and retail trade, finance, insurance, and real estate—have generated nearly 90 percent of the new jobs in New Jersey."

Volvo North America corporate headquarters are located in Rockleigh, New Jersey. Courtesy, Volvo North America Corporation

JAGUAR CARS INC.

The Jaguar executive management team: (left to right) Edward J. McCauley, senior vice-president/finance and administration (seated); Graham W. Whitehead, president, Jaguar Cars Inc. (standing); and Michael H. Dale, senior vice-president/sales and marketing.

The United States is currently home to approximately 300 Jaguar employees working in two facilities based at opposite ends of the country. Established in 1968, Jaguar's U.S. headquarters, in Leonia, New Jersey, 10 miles from the heart of New York City, is 2,600 miles away from Jaguar's Western Zone office in Brisbane, California.

In between the two coasts are the firm's sales and service representatives, working with Jaguar's network of dealers to market approximately one-half of Jaguar's annual production. The company also maintains a test facility in Phoenix, Arizona, where year-round testing of new models is carried out.

Jaguar Cars Inc. will establish its new U.S. headquarters at the Ramapo Ridge-McBride Office and Research Center in Mahwah, New Jersey, in late 1989. The $25-million project, developed by McBride Enterprises, contains a 200,000-square-foot facility designed by the New York architectural and engineering firm of Haines, Lundberg, Waehler. Selected for its centralized location to the eastern metro market, the facility is located on a 20-acre site in the foothills of the Ramapo Mountains on MacArthur Boulevard near Route 17, two miles south of the New York Thruway.

The United States is the world's, and Jaguar's, largest automobile market. Jaguar has carved out a strong position in the luxury car field but faces terrific competition, especially from other luxury cars that have long enjoyed high popularity with the well-to-do car-buying public. According to Jaguar president Graham W. Whitehead, "The appeal of the Jaguar car is its understated British elegance and style. Our cars are both sophisticated and refined in the way they look and drive."

According to senior vice-president/finance and administration Edward J. McCauley, executive in charge of the move to Mahwah, "This New Jersey metro-area location was selected by Jaguar not only for its highly accessible location but also for the strong employee base the area offers."

Since the first Mark IV crossed the Atlantic in 1948, U.S. distribution and marketing of Jaguars has changed significantly. Jaguars are shipped from Coventry to either Port Newark, New Jersey, or Port Los Angeles, via the Panama Canal. The trip to Newark takes seven to eight days, while it takes three weeks, including stops, for a car to reach Los Angeles.

Jaguar's sales grew from just surpassing 3,000 units in 1980 to approximately 25,000 by year-end 1988—the result of close cooperation between Jaguar Cars Ltd. in the United Kingdom, led by Jaguar chairman and chief executive Sir John Egan, and the U.S. marketing team.

According to Egan, "Without the United States market, Jaguar's current strength in the world automotive field would have been nearly impossible. There was a huge untapped pool of potential buyers in the United States. When Jaguar began building a car that American dealers and these potential customers could feel confident about, then, given a dependable supply of cars, it was only a matter of time before sales started to climb."

John Egan was hired as chairman of Jaguar in 1980, at the time Jaguar began operating as a separate unit of British Leyland. In July 1984 Jaguar plc was established as the holding company for the Jaguar Group and shares of Jaguar plc began trading in August 1984 on the London Stock Exchange. American Depositary Receipts (ADRs) representing Jaguar plc common stock are available through U.S. brokerage firms.

The importance of the United States to Jaguar is demonstrated first by the fact that nearly 50 percent of all Jaguar cars are sold here. Second, nearly 30 percent of all Jaguar shares are held in the United States. Jaguars are being sold in great numbers to people who have enough confidence in the company to invest in it as well as purchase its product. This trend is expected to continue, making the bond between Jaguar and the United States even stronger.

New facilities have been opened by many dealers and more are currently

An artist's rendering of the Jaguar Cars Inc. new U.S. headquarters at the Ramapo Ridge-McBride Office and Research Center in Mahwah, New Jersey. The $25-million, 200,000-square-foot facility was selected for its centralized location to the eastern metro market.

under construction as part of Jaguar's $200-million dealer up-grade program which has been initiated to ensure that customers receive the finest possible service. To support dealer service efforts, the New Jersey and California locations maintain comprehensive parts operation to ensure quick response to customers' needs.

Jaguar has actively participated in motor sports in the United States since 1974 and, since 1982, has competed in the International Motor Sports Association Camel GT series with XJR-5 and XJR-7 V12-powered prototypes. According to Michael Dale, senior vice-

A 1988 Castrol/Jaguar XJR-9 racing car.

president/sales and marketing, who directs the racing program, "As of 1988 Jaguar has forged a new racing partnership with the Wayne, New Jersey-based Castrol Company, which will further increase our strength in motor sports competition. Without racing, a Jaguar wouldn't be a Jaguar."

Jaguar Cars Inc.'s support of the United Way and a number of local charities and educational institutions reflects its belief in being a good corporate citizen. Employees who are actively involved in local civic organizations are encouraged to do so by Jaguar.

Jaguar was founded by William Lyons in England in 1922 as the Swallow Sidecar Company. The first car to bear the Jaguar marque was the SS Jaguar 2.7-litre Saloon in 1935. In 1945 the firm became Jaguar Cars Ltd. It merged with British Motor Corporation in 1966, and two years later with Leyland Motor Corporation to form British Leyland Motor Corporation Ltd. Jaguar Cars Ltd. has been operating independently since its privatization in 1984.

In the field of construction, recent statistics indicate that New Jersey growth falls within the top 10 percent of all the United States. Photo by Bob Krist

In the same article, Samuel Ehrenhalt, commissioner of the United States Bureau of Labor Statistics, lauded New Jersey's performance in adding nonmanufacturing jobs:

It has been among the best in the nation. Today, over three-fourths of New Jersey's jobs are service producing, up from 70 percent in 1979. Job growth rates in New Jersey in the 1979-1986 period ranked in the top 10 fastest-growing states in services and construction, and in the top 15 in trade and finance, insurance, and real estate.

Asserting that "the job record for New Jersey in the 1980s is extraordinary for an older industrial state," Ehrenhalt points out that "New Jersey really has geared up to provide formidable competition" for Manhattan.

During the last seven years, New Jersey's economy has been fur-longs ahead of other state economies. The state's record during the 1980s led the revitalization of the entire northeastern portion of the

To accommodate the rise of interest in the service-oriented industries, many new facilities such as this office building in Englewood Cliffs are being constructed, thus creating twice the number of available jobs. Photo by Bob Krist

SHANLEY & FISHER, P.C.

The law firm of Shanley & Fisher combines a tradition of the highest-quality legal counsel established by the firm's senior partners, Bernard M. Shanley and Harold H. Fisher, with the skill and acumen of a group of the state's most modern and forward-thinking legal practitioners.

Founded in Newark in the midst of the Great Depression by Joseph Young and Bernard Shanley, the firm was later joined by Harold H. Fisher. The company took the name of Shanley & Fisher during the 1950s.

Both attorneys were well known for public service. Fisher served as special assistant to prosecutor David Wilentz during the trial of Bruno Hauptmann for the kidnapping and slaying of the Lindbergh baby. Shanley became appointments secretary and personal counsel to President Dwight D. Eisenhower during his tenure, and was also a candidate for United States senator and New Jersey governor.

Building on this foundation of service to and knowledge of the Garden State,

Shanley & Fisher has developed into a broad-based, full-service practice with specialization in such areas of law as product liability, taxation, real estate, trusts and estates, corporate, and the growing field of environmental law.

The firm's professional staff numbers some 100 attorneys, of whom 36 are partners, including a former attorney general of the State of New Jersey, and other former and current elective officials at the state and local bar associations, and many teach at law schools throughout the region. Assisting the professional staff are 180 paralegal, secretarial, and other support personnel.

The firm opened its Morristown headquarters at 131 Madison Avenue in 1984. In addition, Shanley & Fisher has

Shanley & Fisher's headquarters is strategically located on Madison Avenue in Morristown. The firm specializes in product liability, taxation, real estate, trusts and estates, corporate law, and environmental law.

full-service satellite offices in Somerville and New York City, both opened in 1987.

As a further indication of its innovativeness, Shanley & Fisher established a unique subsidiary, Issues Management, Inc., in 1987. This operation, based at the Carnegie Center in Princeton, broadens the group's capabilities in such areas as strategic communications and public affairs, and enables it to help solve problems in areas not suited to traditional approaches.

New Jersey has emerged as one of the most sophisticated states in the union in terms of its economic, political, and social issues. It is certainly a trendsetter in many applications of law, to both regional and national issues.

Shanley & Fisher has grown to meet new and ever-changing challenges for its clients. The firm is committed to providing aggressive, innovative legal services for businesses, institutions, government agencies, and individuals within the state, the region, and nationwide.

United States, and shows no signs of flagging, even in the face of the national calamity experienced on "Black Monday" in October 1987. During the 1980s, 500,000 new jobs were generated by the state's healthy economy, and it is projected that 600,000 more new jobs will be created in the next five years.

As more and more corporations discover northern New Jersey's agreeable business environment, new jobs continue to be created. The region's phenomenal success, in turn, has bolstered the state to assume its present status as the Northeast's economic wonder. With every passing year, northern New Jersey's vigorous economy helps to add another chapter to New Jersey's "success story."

A SPECIAL CLIMATE FOR FREE ENTERPRISE   In order to continue the healthy environment which has nurtured the phenomenal growth in northern New Jersey, the Commerce and Industry Association of New Jersey has been instrumental in bringing businesspeople together in formal and informal settings. As a member of the Association and a past chairman of its board of directors, J. Fletcher Creamer, president of J. Fletcher Creamer & Son, Inc., describes the advantages available to members: "I believe it provides excellent opportunities for meeting businesspeople and sharing problems and solutions that work. I've been quite active in it, and I've enjoyed it. For example," he continued, "there have been times in a meeting when a businessperson has said something quite simple which has resulted in helping me solve a current dilemma."

The association accepts all businesspeople, and the mix is interesting and stimulating. "Everyone has something to offer," said Creamer. "In the end, we all have the same inherent problems …taxes, infrastructure, payroll, employment, etc. We all have common problems, plus, each industry has its particular set of problems. As a result, we learn from each other."

In discussing the association's importance to the business community, Creamer said:

It keeps busy business people abreast of what the legislators are doing in Trenton. They track many of these so called anti-business bills and put us on record so that we can write to our legislators. I think the Association is invaluable in creating and maintaining a healthy interest in our economy and New Jersey's success.

As a strong advocate of education, the Commerce and Industry Association has also been interested in teaching future leaders about how and why the free market works. It accomplishes that through its tax-exempt non-profit affiliate: the Foundation For Free Enterprise. Jim Cowen, Association president, and Chip Hallock, executive vice-president, point out the merits of a free enterprise system: "It allows people to choose and do what they want to do, free from restriction and government dictates," Cowen said. "People respond to incentives, and what is in their own best interest. That is why the system works, and it is also why so many other countries are now implementing facets of capitalism and private enterprise into their own systems—like England and China." In describing the foundation's work, Cowen stressed that "it is completely non-political and non-partisan, and it does not discriminate as to who can participate. It's a completely economic/educational endeavor." In outlining the foundation's mission, Hallock said, "we try to bring groups into a classroom and teach

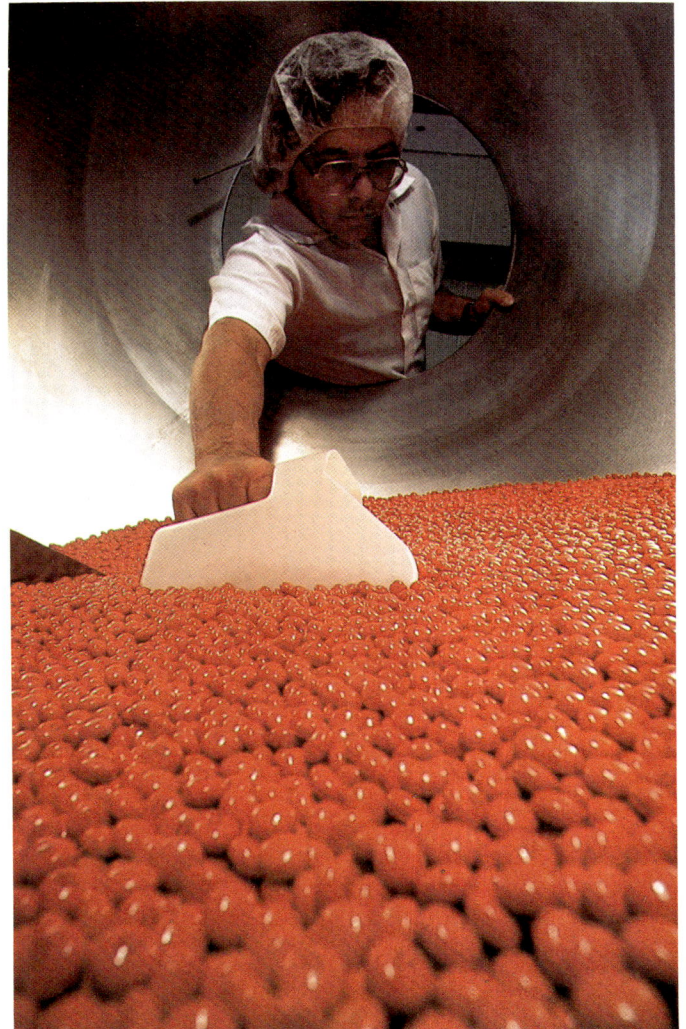

The work force available to New Jersey employers today is both highly educated and highly skilled. Photo by Bob Krist

them the basics: supply and demand, inflation, free trade, and the minimum wage and its effects." Cowen underscored the importance of education in these areas because students leave the current educational system largely ignorant of how the system operates.

We feel that they do not get this education anywhere else. Most high schools don't offer an economics course. Students never get to learn about the intricacies of the free enterprise system and the beauty of its structure and success. We hope to make them understand that less government intervention is better for the economy as people more freely decide what they want to pay for. We help them to understand the effects taxation has on the economy. By doing this, we feel that we help to create a better informed citizen.

The Foundation For Free Enterprise is supported by contributions from companies, foundations, and individuals. According to Hallock, "These groups and individuals are similarly committed to advancing the understanding of the free market."

Northern New Jersey is currently experiencing a major construction boom. Photo by Michael Melford

It's a piece of New Jersey where real estate projects are measured in the billions of dollars. It's a place where parcels of land inspire names like "the Gold Coast," "The Meadowlands," and "International Crossroads." It's big scale regentrification and innovative new construction. It's the state's dynamic northeastern region, and it's a developer's dream come true.

The proverbial "three words" to describe real estate value—"location, location, location," —must have been coined in northern New Jersey because of the area's obvious and fortunate geographical position. Its proximity to major markets, its healthy economy, and its extensive network of highways and transportation systems combine to

CHAPTER THREE

make it an unbeatable competitor. All these factors have served the real estate industry well, and throughout the region there is strong evidence of a healthy and vigorous construction expansion.

THE GOLD COAST Dubbed the "Gold Coast," the western coastline of the Hudson River lives up to its name. With its vast number of major new real estate projects and the programs of revitalization within

A Developer's Dream

Above: The landmark Colgate Clock was installed atop an eight story building at the Colgate-Palmolive plant in Jersey City in 1924. The face of the clock measures 50 feet in diameter and the minute hand moves 31 inches-per-minute. Photo by David Greenfield

the established cities that make up the area, it has become one of the most exciting harborside developments on the East Coast.

Among the newest plans to be unveiled involves the redevelopment of the landmark waterfront Colgate Palmolive Company in Jersey City. The elaborate plan calls for six million square feet of office space, a 400-room hotel, 1,500 residential units, 250,000 square feet of retail space, and a 400-slip marina. The estimated cost will exceed $2 billion. The project's first phase, which will be ready for occupancy in three years, is a 1-million-square-foot office building. According to Reuben Marks, quoted in an article in the *Star Ledger,* the project will provide 1,600 construction jobs over the 15-year life of the plan.

Newport is perhaps the most ambitious project being attempted along the waterfront. The mini-city is the largest single urban develop-

Right: One phase of the Newport Centre redevelopment project in Jersey City includes this Newport Yacht Club. Photo by Carol Kitman

ment project in the nation. Encompassing 560 acres, this abandoned railroad yard is being transformed into a spectacular mixed-use complex (residential and commercial) at a cost of over $10 billion. What was once a developer's nightmare promises to become a dream city along the Hudson, overlooking the world's most recognizable and spectacular skyline.

The first phase, Newport Centre, which opened in October, is the project's tri-level enclosed regional shopping mall—Hudson County's first major mall. It contains 1.2 million square feet with four major department stores (Stern's, Sears, Macy's, and a fourth to be announced), 165 specialty stores, and a nine-theater cinema, plus restaurants.

In addition to Newport Centre, Newport will feature 15,000 units

(1,500 rental, 13,500 condo); 4 million square feet of office space; hotels; New Port Yacht Club and a 1,000-slip marina; a 36-acre park; an oceanographic center/aquarium; and theaters and restaurants. Access to the new "city" will be easy. The PATH train stop is at the center of the region, with the Holland Tunnel at its doorstep. There will be ferries to six Manhattan stops, water taxis, a seaplane base, and a helipad. The LeFrak Organization of New York will build the office and residential towers and Melvin Simon and Associates of Indianapolis is developing the regional mall.

Other major projects along the waterfront have also contributed to its growing reputation as a "priceless" piece of real estate. Its obvious and tantalizing advantages have lured imaginative developers to create a new "El Dorado."

Harborside, a $900-million financial center, is a mixed-use development in Jersey City. It will contain six million square feet of office space, a 70,000-square-foot retail center, a river promenade, waterfront residences, a 250-slip marina, residential towers, hotels, and restaurants. The project is adjacent to Newport. It is estimated that the 48-acre project will be completed within a 10-year period, creating 450 construction jobs.

North of Harborside, Harsimus Cove adds itself to the list of innovative mixed-use complexes. Directly across the Hudson River from the World Trade Center, the project will comprise 2 million square feet of offices, 450 townhouses, and 1,750 highrise residential units, plus a marina, restaurant, and retail area.

Exchange Place Centre is a 30-story, 700,000-square-foot office tower. First Jersey National Bank, a co-developer, will take the top seven floors of the skyscraper for its new corporate headquarters. The other joint venture partners are William D. Shaffel and The Prospect Company of The Travelers. The 490-foot green-glass structure is the tallest building in the state. Its distinctive design is the result of the cre-

The Harborside Financial Center is a $900 million mixed-use project in Jersey City. Photo by Carol Kitman

Upon completion, the Exchange Place Center in New Jersey City will be the tallest building in the state. Photo by Carol Kitman

ative efforts of the Grad Partnership.

Just a short distance west of Exchange Place Centre, the new International Financial Center is being created by Cali Associates in partnership with James Demetrakis. The new 19-story, 600,000-square-foot edifice will be one of the largest office buildings ever built in Jersey City. It was designed by New York architects Herbert Beckhard Frank Richland and Associates. Half of the office space has already been leased to The Pershing Division of Donaldson, Lufkin and Jenrette, a major brokerage firm. They will move into the building in the spring of 1989.

Port Liberté, a $600-million luxury residential development, adjoins Liberty State Park. Already being called the "Venice of the Northeast," the mixed-use development will also feature shopping facilities, a world-class hotel, several restaurants, a yacht club, and various recreational facilities.

Evertrust, a $25-million, 17-story office building, has already been completed, and a second 28-story tower is planned.

A Hartz Mountain Industries joint venture will develop Journal Square Corporate Center. The development will include four office buildings totaling 1.6 million square feet adjoining the PATH transportation center. A pedestrian skywalk will create the linkage from the Journal Square Corporate Center to the transportation facilities. The Exchange Plaza PATH terminal is undergoing a $51-million renovation to keep pace with the area's rapid growth.

The development in Jersey City has not ignored the need for affordable housing, and developments are afoot to meet those needs. The former Joseph Dixon pencil factory is being converted into more than 500 affordable residential apartments at a cost of $30 million. The abandoned Betz Brewery building is being rehabilitated and converted into low- and moderate-priced rental units.

Commenting on the area's development, John Cali, a prominent real estate developer, said that New Jersey is becoming an expansion of Manhattan. "For a four minute commute," he said, "firms are finding lower taxes, more attractive rents and utilities, and a vastly improved quality of life compared to Manhattan."

The redevelopment phenomenon of Jersey City has attracted developers from Ohio and Massachusetts and from as far away as Taiwan, and the explosive growth shows no signs of slowing. When

tallied, the combined figures for square footage are staggeringly impressive. Rick Cowen, director of Jersey City Housing and Economic Development, reports the results: "We've added up the total of all the projects now under construction or recently built in Jersey City, and we've come to some impressive figures," he said.

Between now and the year 2000, we expect to generate 26,915 dwelling units, 25.8 million square feet of office space, more than two million square feet of retail shopping, 1,600 hotel rooms, and 2,783 boat slips. Obviously, it will have a sizable impact not only on Jersey City, but the whole region. In Jersey City we will have the addition of 60,000 new residents by the year 2000. We have potentially 70,000 to 100,000 new jobs and the potential of capturing all those retail dollars that have leaked out of the city and out of the state.

BERGEN'S BRILLIANT SUCCESSES   In northern New Jersey, Bergen, Passaic, Morris, and Essex counties are also undergoing growth in real estate development. According to the *New York Times* (June 9, 1987), "Twice as much office space is under construction in New Jersey as in New York or any other market in the nation." To further support the market strength in these counties, *The Record,*

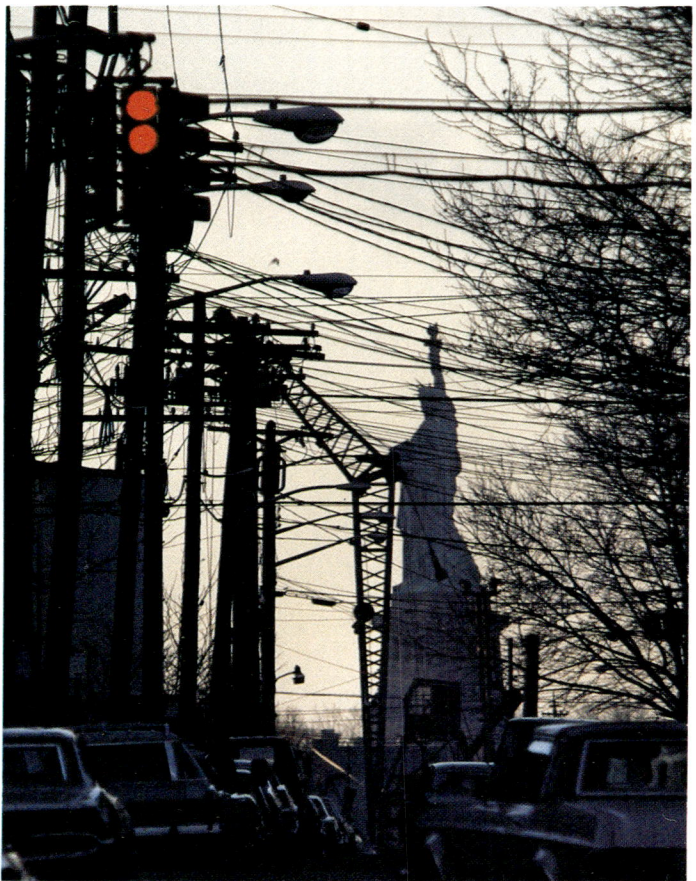

Above: Projections by the director of Jersey City Housing and Economic Development indicate that by the year 2000, Jersey City will have generated 70,000 to 100,000 new jobs, attracting 60,000 new residents. Photo by Bob Krist

Right: The juxtaposition of high-rise commercial-residential space with private homes is becoming more and more commonplace as major new mixed-use complexes are currently under construction across northern New Jersey. Photo by Bob Krist

BELGIOVINE CONSTRUCTION GROUP

There is no greater testament to the American dream of building one's own business through individual initiative and determination than the success of Giuseppe Belgiovine, founder and president of the Belgiovine Construction Group of Hackensack.

This native of Italy, who emigrated to the United States in the 1960s, set a course for success upon his arrival, and today—with his sons, Frank and Vincent—operates a construction and real estate development company renowned for its high standards of quality and service.

Giuseppe Belgiovine has been involved in construction since he was a 16 year old in his native land, where he became a construction manager prior to coming to the United States. Despite his early success, he knew that his ambition could not be satisfied without coming to the land of opportunity. His wife's family had already emigrated to the United States, and they urged Giuseppe to follow. Belgiovine, his wife, and their young children did so in 1966, arriving in Hoboken where thousands of his countrymen first lived.

Belgiovine had to start over as a laborer and mason in his new country. His hard work and dedication to quality earned him rapid promotions, but it was not enough: He wanted his own business. At first he attracted small side jobs

Another Belgiovine project, the Stern's Department Store, anchors the Newport Centre Mall in Jersey City.

The Belgiovine Construction Group has handled a number of projects for The Record, one of which was this major addition to its headquarters in Hackensack.

while working full time for others. Primarily he did masonry work such as patios, sidewalks, and porches—projects that could be done at night and on weekends. Then, with the help of a friend who cosigned a loan for an old dumptruck, Belgiovine set out on his own for good.

Initially the Belgiovine Construction Group serviced the residential market by doing home renovations and remodeling. Hampered at the time by his unfamiliarity with English, Belgiovine relied on his younger son, Frank, to translate during conversations with clients and to assist in writing proposals. Frank, now a vice-president in the family business, was not yet 10 years old at

the time.

With the advent of the 1970s Belgiovine began building homes, first in Secaucus, where the building boom that engulfed the entire Meadowlands region began. The high quality of Belgiovine's work became well known, and the demands for the contractor's services snowballed. The company was responsible for the masonry work at a town house development under construction in Englewood, while also undertaking residential renovations

throughout Hudson and Bergen counties. In 1976 Belgiovine Construction built its first office building (in Edison), and soon thereafter made a commitment to building large commercial and public works projects.

During the period from 1978 to 1983, the bulk of the company's business was in serving the public sector. Beginning with the board of education building in West New York in 1978, Belgiovine knew he had set a different course of direction for his company with additional projects such as The Lyndhurst Health Center, the Morristown Police Station, veterans' facilities, schools, the Military Ocean Terminal in Bayonne, and the renovation of the Hoboken Boys Club.

Riding the crest of the firm's reputation for service and quality, Belgiovine opened a Lyndhurst office in 1978 to reach the burgeoning construction markets of the Meadowlands and beyond into Bergen, Passaic, Essex, and Morris counties. Soon some of the firm's earliest clients were calling on Belgiovine for increasingly larger and more complex projects, including the construction of an entire industrial park in Orange, New Jersey, in 1982.

Belgiovine's crews and equipment were expanding throughout New Jersey from the New York border on into Monmouth and Hunterdon counties. The company's growth period actually exploded about four years ago when in quick succession came The Record's regional headquarters on Route 23 in Wayne, the Mulberry Street Market in Newark, a major high-rise apartment building in Guttenberg, additions and renovations to The Record's Hackensack headquarters, and work at the newspaper's affiliate, The News Tribune, in Woodbridge.

Other projects completed are the Kids 'R' Us store in Jersey City, a major addition and renovation to the Hudson Mall in Jersey City, the Passaic County Juvenile Detention Center in Haledon,

Belgiovine Construction Group also built the Hudson Mall in Jersey City.

Stern's Department Store at the Newport Centre Mall in Jersey City, a five-story addition to the Quality Inn in Hasbrouck Heights, the General Aviation (Industrial and Office) Building in Hackensack, and Belgiovine's own corporate headquarters building, also located in Hackensack.

Future assignments include a regional headquarters building in Englewood for National Community Bank, as well as an administration building for the East Orange Water Department, an education center at the Frelinghuyser Arboretum in Morristown, and a 150-unit luxury condominium building in Fairview.

These and other projects demonstrate the skill and diversity of the Belgiovine Construction Group, now the building arm of Belgiovine Enterprises, Inc. The firm does all of its own excavation, concrete, and masonry work, as well as carpentry, while subcontracting the mechanical and electrical trades, road work, and finishes. Belgiovine permanently employs some 100 skilled workers in the building trades.

Construction is not the only concern of the Belgiovine Construction Group; the company has become increasingly involved in real estate development,

both independently and through joint ventures. Belgiovine plans to develop five acres in Union City for a 1,200-unit condominium project and 200 additional apartments at a different site less than two miles away, as well as an office building in Hackensack. The firm is also expanding westward with properties in Sussex County.

Vincent and Frank continue to be key components in the growth and diversification of the company. Both worked with their father while attending school, joined him in the field on projects, operated machinery, drove trucks, helped in the office, and became deeply involved in the sales efforts directed toward new business development.

Complementing and augmenting the skills and vision of Giuseppe Belgiovine and his sons, the firm has actively sought top professionals in construction and finance, such as controller Dominic Mustillo from the national accounting firm of Touche Ross, as well as architects, engineers, project managers, and legal and administrative staff.

The Belgiovine Construction Group, headquartered at 325 South River Street in Hackensack, is prepared for the challenges and opportunities of the future, based on the high standard of quality service that has brought the company its success.

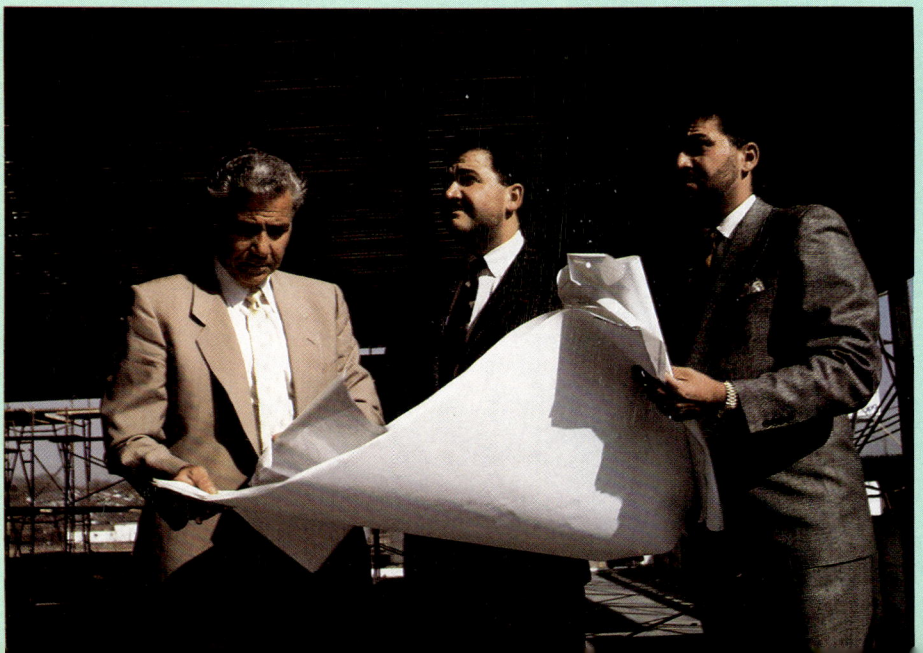
Giuseppe Belgiovine (left), founder of the Belgiovine Construction Group, and his sons, Frank (center) and Vincent (right), discuss the expansion of their Hackensack headquarters.

J. FLETCHER CREAMER & SON, INC.

The dramatic development of northern New Jersey during the past half-century has depended on the presence of full-service contractors to build structures, roads, pipelines, underground utilities, and bring projects to fruition.

J. Fletcher Creamer & Son, Inc., founded in 1924, is the archetype of the diversified contracting industry, serving businesses, governments and agencies, and private and public institutions throughout New Jersey.

Headed by J. Fletcher Creamer, grandson of the company's namesake and the third generation in the family to bear the name, the firm has grown in scope and size from a fledgling operation whose sole asset was a solid-tire dump truck. Today the company boasts a fleet of vehicles, as well as the most modern, sophisticated excavation and construction equipment available.

But the key to Creamer's success lies not only in its hardware, but also in the skill, motivation, and dedication of the approximately 500 workers employed there. With the close of the twentieth century approaching, the firm's commitment to value and quality is as important now as it was more than 60 years ago.

In addition to the company's modern equipment and sizable work force, J. Fletcher Creamer & Son is significantly different from the early days of its progenitor. That difference lies in the tremendous diversity of the company's related lines of business.

From the beginning J. Fletcher Creamer (the first) and his son had sought to broaden their business base. At first their old Ford truck was used for odd jobs and deliveries. Then they entered into the coal and coke business, while also servicing local contractors for on-site delivery and removal of fill. Later they began a fuel oil business.

It was not until the 1960s that the firm concentrated solely on construction and related businesses. And it has been in the field of construction that the company has left an indelible mark. Radiating from its original base in Fort Lee—the operation later moved to Edgewater and now has its headquarters in Hackensack—Creamer took on in-creasingly large and important assignments.

During the 1930s the firm hauled rock and soil from the excavations leading to the construction of the George Washington Bridge. This may have been the single most important develop-ment leading to North Jersey's current rise as one of the most attractive and convenient settings for corporate, commercial, and residential growth in the United States, if not the world.

The company excavated foundations from the hard rock of the Palisades for

A J. Fletcher Creamer & Son, Inc., project was the Bergen County Utility Authority Project Baler Facility in North Arlington, New Jersey.

modern society and its economic base depend. In this regard, the firm performs construction services for major utilities in New Jersey, including Hackensack Water, New Jersey Bell, Orange & Rockland Utilities, PSE&G, plus AT&T in New York and Pennsylvania, as well as in New Jersey.

The variety of Creamer projects is awe inspiring. From pipework of all kinds (gas, water, sewer lines, telephone, and television cables from Florida to Kentucky and up to the Canadian border) to highway and bridge projects for the New Jersey Turnpike, Garden State Parkway, the Port Authority of New York and New Jersey, and the New York Department of Transportation, teams of Creamer experts are on the job year round.

Each day finds the company's own helicopter, acquired in 1986, logging hundreds of miles transporting project managers, specialists, and (as needed) vital replacement parts from corporate headquarters on East Broadway in Hackensack to widely scattered construction sites. But like the darting chopper that connects the organization's widespread projects, J. Fletcher Creamer & Son continues to move decisively as a business enterprise.

The Creamer organization is taking on highly technical installations of fiber-optic lines for AT&T, MCI, and US Sprint, while also making a major effort in cleaning and relining water mains that will have an important role in extending the life and usefulness of these vital infrastructure links. And, allied with its highway construction operations, subsidiary companies manufacture and market road signs, guardrails, and other related items.

With a firm foundation from the past excavated in the rock of the Palisades, J. Fletcher Creamer & Son Inc., is growing toward the future. At the helm is chairman J. Fletcher Creamer. The day-to-day operations are handled by Fletch Jr., president, ably assisted by his brothers, Glenn, Dale, and Jeff.

most of the high-rise structures forming the skyline on the New Jersey shore of the Hudson River, after having erected and later demolished the exciting rides of the famed Palisades Amusement Park. Creamer is now integrally involved in the construction and development of projects throughout the Garden State, as well as into New York, Connecticut, and other states.

Rated among the nation's top 250 contractors, J. Fletcher Creamer & Son has as a major focus of its business the infrastructure facilities upon which

The Courts of Glenpointe offer lux-ury-equipped 1, 2, and 3 bedroom townhomes, round the clock secu-rity, and a private health club. Photo by Carol Kitman

Bergen's largest newspaper, reported that of the 13 million square feet of office space leased in the state's 12 northern counties, Bergen, Passaic, Morris, and Essex counties accounted for half.

Among Bergen County's premier real estate projects is Glenpointe, a 50-acre, mixed-use development in Teaneck. Glenpointe has strong and obvious location advantages. It is only three miles west of the George Washington Bridge, at the intersection of Interstate 80 and the New Jersey Turnpike—one of the most important crossroads in the entire New York Metropolitan area.

The focal point of the complex is the 350-room Loew's Glenpointe Hotel and Conference Center. Also of special note are the 22,000-square-foot Glenpointe health club/spa, a retail mall, two of-fice buildings totaling 567,000 square feet, and a luxury residential phase called The Courts of Glenpointe.

Glenpointe represents an exciting achievement for Alfred Sanzari Enterprises, a 40-year-old firm. The firm, which was founded by Al-fred Sanzari, the present chairman, has grown along with the area in which it has carried out some of its most ambitious projects. During the postwar baby boom, Sanzari built and marketed approximately 3,500 one- and two-family homes, primarily in Bergen County. Within 10 years, the firm was ready to move into the larger challenge of con-structing garden apartment complexes throughout New Jersey. Many of Sanzari's commercial and industrial properties in South Hackensack, Norwood, Elmwood Park, Little Ferry, and other Meadowland communities are still owned by the Sanzari organization.

Since the mid-seventies, the firm has become highly active in New Jersey's dynamic commercial real estate market, helping meet the demand for new office space. Two early projects for this market were Heights Plaza in Hasbrouck Heights on Route 17 overlooking I-80, a 200,000-square-foot, six-story office building; and Elmwood Park Plaza, a 64,000-square-foot office building close to I-80 and the Gar-den State Parkway. The practical experience gained in these projects, which were completed in 1980, laid the groundwork for Glenpointe and its successor: Court Plaza North and South in downtown Hacken-sack. Commenting on the project in *Real Estate Forum*, David Sanzari, president, outlined their reasoning for attempting the project:

Indeed we still remember very vividly that many people were mystified by our confidence in downtown Hackensack in the early 1980s when we began developing Court Plaza North, at a time when many businesses were leaving this city for greener suburban pastures. But the downtown districts of our country have tremendous natural resources, which, when combined with well-planned, quality-oriented development, can be a very potent combination.

Chairman Alfred Sanzari is still an active member of Alfred Sanzari Enterprises, but in 1984 he turned over the day-to-day man-agement to his son David, now president and chief executive officer.

THE MEADOWLANDS MIRACLE   The Meadowlands Sports Complex has helped commercial development growth in the area. Greatly improved highways, built to serve the three-part sports mecca, were the catalyst for new construction of office parks, hotels, and major distribution centers.

Within this area, one of the largest multifaceted projects is the vast complex developed by Hartz Mountain Industries. In the *New York Times* (August 8, 1987), Hartz's visionary Chairman Leonard Stern described the moment that he learned the Meadowlands area was for sale: "I thought I had seen the next coming. People told me I was crazy when I wanted to buy the land in 1968, but I bought it anyway." That foresight paid off. The area was transformed from a swampy garbage-laden property to one of northern New Jersey's swankiest business addresses.

Today the Hartz organization has leased over 25-million square feet of office and industrial space and has constructed 1,200 residential units. The new Harmon Meadow project, a unique 500-acre commu-nity in the Meadowlands, has 3.5-million square feet of office space plus shopping, hotels, homes, and entertainment facilities. In Ridgefield, Hartz will build one million square feet of office space, a hotel, and retail space, surrounded by a park. Besides its properties in

Glenpointe Conference Center main-tains exceptionally well-equipped small and large conference rooms with audio-visual capacities. Photo by Carol Kitman

More housing developments such as this one near Freehold will be required to accommodate the anticipated influx of new residents. Photo by Bob Krist

the Secaucus area, Hartz is developing other sites in the northern New Jersey region, including areas on the Hudson waterfront and in more than a dozen northern New Jersey communities.

In the northwestern area of Bergen County, J.D. Construction Corporation is developing the $300-million International Crossroads in Mahwah. The spectacular landmark complex will have no peer: it is a first in its area. But then, J.D. Construction is accustomed to firsts— in its 60-year history, the firm's successes have placed it in line with New Jersey's best.

International Crossroads is the transformation of what was once the site of the nation's largest automobile production plant housed under one roof: the 2.3-million-square-foot Ford Motor Company assembly plant, which closed its doors in 1981. J.D. Construction put the property, which had lain fallow, to good use. They sold a 65-acre parcel to Sharp Electronics for its North American headquarters, and retained 107 acres for other uses. The developer could have utilized the building for warehouse space, but instead elected to pursue a much more ambitious plan. With a staunch commitment to the project, it has commenced to transform a virtually useless, abandoned site into a classic corporate complex.

J.D. Construction spent $5 million to clear the site for the first phase of construction, which involved the complex's centerpiece: a 22-story hotel/office building comprising 550,000 square feet, which offers 370,000 square feet of office space and 180,000 square feet of hotel space. The Sheraton Crossroads Hotel and Convention Center will occupy nine upper floors of the building, and it will contain 230 deluxe guest rooms. The ground level will contain restaurants, lounges, banquet and meeting rooms, a health club, and an entertainment center.

The new complex's amenities include on-site independent electricity. In an article in *Commerce Magazine*, James D'Agostino, Jr., president of the firm and son of its founder and chairman, said, "There will be a dual source of power, which will offer tenants 100 percent

independence from local utility service in case of downtime. This makes the site a natural for a computer operation, for it offers a failsafe system. Moreover, tenants will not have to invest in supplementary air conditioning." The complex will also offer the convenience of fiber optics and a sophisticated shared-use telecommunications and data service on-site, along with the latest innovations in electronic building management, energy efficiency, and security.

International Crossroads is aptly named because it is literally at the center of the crossroads of the tristate area, located at the intersection of the New York Thruway, I-287 (to be completed by 1989), Route 17, and Route 202, with connections to I-80, the Garden State Parkway, and Palisades Parkway among others. It is close to New York City; it is in the center of the BoWash (Boston to Washington) Corridor, and it lies midway between Morristown and White Plains along I-287. James D'Agostino, Sr., chairman of the company, commented on the firm's intentions in *Commerce Magazine:* "We are creating an international theme here at the corporate park, and we believe a number of tenants will be firms with international operations." This theme will be reflected in the complex's aesthetic landscaping, with gardens in French, Oriental, Baroque, and English motifs. "We are creating a showplace for the state," D'Agostino said.

Upon completion, International Crossroads will be the apex of a 60-year tradition of accomplishment for J.D. Construction. Garnering numerous awards over the years, the firm was responsible for such notable buildings as Stonehurst at Tenafly, a 250-home community of single homes on a rocky bluff in Bergen County. It also built its namesake Stonehurst at Freehold, which included 1,000 apartments and 700 homes. The firm was responsible for bringing the first high-rise office building to Bergen County, in 1969. This project became the showplace three-building Continental Towers at Route 4 and Hackensack Avenue in Hackensack. The firm was also responsible for The Camelot, a 21-floor luxury building that was the first high-rise residential condominium in Hackensack. They have completed the modernistic 68,000-square-foot office showcase in Clifton, which served as corporate headquarters for the firm before it moved into the International Crossroads in Mahwah. With a long list of architectural credits to its name, J.D. Construction continues as one of northern New Jersey's most creative developers helping to shape the state's future.

Specialty stores are numerous within the retail marketplace of northern New Jersey. Photo by Dick Luria

When the Garden State Plaza opened its doors in May 1957, it was instantly proclaimed northern New Jersey's merchandising marvel. At a time when the shopping center industry was in its infancy, the center quickly drew the attention of the curious, who came from all over the state to sneak a peak at what the future promised. Only six months after that event, the area surpassed itself by adding another major regional mall less than one mile from the first. When Allied Stores Corporation opened the Bergen Mall, northern New Jersey's reputation as a mecca for shoppers was firmly established. Since those early days of shopping center history, northern New Jersey has continued to maintain its place of prominence as a bona fide retail phenomenon.

CHAPTER FOUR

Today, the most beautiful and complete regional malls thrive within minutes of each other. These luxurious leviathons sprawl against northern New Jersey's suburban landscape, offering every possible mercantile item or service immaginable. In Bergen and Passaic counties, eight major malls cater to the consumer, and in Bergen County alone, six major malls coexist within a three-to-four-mile radius. According to Sales And Marketing Management's 1987 Survey of Buying Power, the total retail sales for Bergen County was nearly $8

The Retail Phenomenon

Above: Riverside Square Mall attracts affluent shoppers. Photo by Rich Zila

Below: This fountain setting provides a quiet repose for weary shoppers at the Woodbridge Center. Photo by David Greenfield

billion, and Passaic County amassed retail sales of over $3 billion during the same period. These two counties alone rank eighth in the nation in percentage of households with an effective buying income of $50,000 or more, and they rank twelfth in the nation in per household retail sales. All these figures add up to one of the most formidable retail markets in the Northeast. According to Elizabeth A. Napoli, general manager of Riverside Square mall in Hackensack,

If you look at the demographics of Bergen County, they are extremely strong. That is probably how we got six major centers in such close proximity to each other. There is a very affluent shopper here, with a lot of discretionary income, and when you combine those aspects, stores want to be here. No matter how much competition there is here, there always seems room for more.

Napoli points out that as more massive retail units develop here, the area becomes a destination point for an even larger region:

You know, Bergen County is not shopped by only county residents. There are people that drive from Manhattan, Upper West Chester, and farther, because they know they can have a 'true' shopping day with so many major stores within minutes of each other. In fact, 17 percent of our shoppers are from New York. As far as retailers are concerned, Bergen is definitely the place to be for retailers because they can generate good numbers here.

The bulk of Bergen County's development can be found along the Route 4/17 corridor in Paramus and Hackensack. This area has enticed nearly every major New York metro department store and mass merchandiser to seek a spot in its vital marketplace. These stores include

Bloomingdale's is one anchor store of the Riverside Square Mall. Photo by Rich Zila

Hahne's is a retail leader in New Jersey. Photo by David Greenfield

Bloomingdale's, Saks, Macy's, Abraham & Straus, J.C. Penney, Lord & Taylor, Sears, B Altman, Stern's, Hahne's, and many more. Several of these maintain one or more units within the area. In addition to the large regional malls, the area possesses a number of lively shopping centers plus attractive and unique central business districts in individual towns. The astute shopper can also discover a shopper's "Eden" of high-quality outlet stores clustered like jewels in Secaucus in Hudson County. All things considered, it could easily be maintained that the area encompassing Bergen and its neighboring northern counties is one of the nation's prime retail areas.

In terms of impact as an industry, shopping centers now account for 54 percent of all non-automotive retail sales in the United States, according to Donald L. Pendley, director of public relations for the International Council of Shopping Centers (ICSC). "That 54 percent translates into $554 billion of non-automotive retail sales a year," Pendley says. ICSC, which was established in 1957, is the trade association of the shopping center industry. Its 23,000 members include owners, developers, managers, retailers, and lending institutions involved in shopping centers. During the course of each year ICSC holds meetings and conventions which enhance the exchange of information. Its research efforts provide the industry with information about shopping centers, and its monthly magazine, *Shopping Center Today,* keeps people within the industry abreast of change. According to Pendley, there are 28,500 centers across the land, and of those, 10

OLSTEN SERVICES OF NORTHERN NEW JERSEY

John F. Luneski, president of Olsten Services of northern New Jersey—the eight-office franchise of the nationwide temporary services network—recognized a need more than 25 years ago and has prospered greatly serving that need since.

Formerly an office equipment supplier who sold businesses such items as calculators, word processors, and integrated data processing equipment and the like, Luneski saw a growing need for temporary, skilled office and other service personnel. Seeking to tap the already blossoming northern New Jersey market during the early 1960s, Luneski joined the network of temporary services agencies being established throughout the country by Bill Olsten, entrepreneur extraordinaire.

With his first small, one-person office, opened in March 1964 at 50 Main Street in Hackensack, Luneski commenced a progression that has seen his string of locations spread throughout Bergen, Hudson, and Passaic counties. Headquarters is a modern, three-story office building at the intersection of Mercer and State streets in the heart of the rejuvenated county seat of Bergen County.

Luneski was riding the crest of a wave of business expansion in the northern

An Olsten warehouseman at work, serving the needs of the light-industrial labor market.

part of the state at the moment when thousands of women were looking to return to work. They were seeking temporary, flexible jobs allowing them to reestablish skills while giving them the opportunity to care for their children.

With the tremendous growth of office facilities and service companies in Bergen County and surrounding markets, Luneski's office workers were in demand. Soon he was opening other

Olsten furnishes a variety of temps with office skills, such as this word processor.

facilities in Paramus, Clifton, Wayne, and later Englewood, Secaucus, Fort Lee, and Ramsey.

Expansion of service industries and the technological explosion of the 1970s and 1980s presented new challenges and opportunities for Luneski and his Olsten franchise. It was no longer necessary or economically feasible to employ someone full time for a less-than-full-time need. Many firms began using temporaries for those hours that they would normally have to pay overtime, while also relieving permanent staff from long and tiring hours that often result in low productivity and errors.

"Temporaries save money," explains the veteran employment executive, "by avoiding overstaffing and keeping their permanent personnel free of increased pressure detrimental to morale at a time when the employer needs teamwork the most. Moreover, employers do not have to put temporaries on the books for health insurance, vacations, overtime, pensions, and a host of fringe benefits that may account for roughly half of salaries. In addition, employers are then able to better utilize their fixed assets, equipment (such as computers, word processors, etc.), and facilities."

Entirely new areas have opened up for the temporary services provider. Prominent among them is nursing care, from nurses caring for individuals at patients' homes to the broader applications of skilled temporary nursing and other medical service providers in hospitals, nursing homes, and other facilities.

Luneski had to look no further than his wife, Irene, a registered nurse, to head up Olsten's Health Care Division, based in the Hackensack headquarters. This division supplies registered nurses, licensed practical nurses, and certified health aides for home care, as well as staff relief for hospitals and nursing homes. A unique feature of this service is that Olsten handles the liability insurance, workers' compensation, malpractice insurance, and proper credentials for the R.N., L.P.N., and home health aide. "Our professional health services not only provide the care or staffing help the client needs, but also protect them from the insurance point of view," says Luneski.

With the changing, broadening market, Luneski's organization made adjustments to better serve clients. Chief among these changes has been providing wide-ranging training options for temporary workers registered with the service. "There have been such profound changes in the office environment that it behooved us to provide training to upgrade our applicants' skills in order for them to perform well in client locations," Luneski notes.

Key among the equipment skills being taught are those for word processors and personal computers. "Those coming back into the labor market after a prolonged absence due to child care or other reasons find that they are not current in various types of office equipment," he says, "while others working with us may also want to expand their familiarity with different kinds of equipment and acquire new skills. This certainly makes them—and us—more marketable."

R.N.s, L.P.N.s, and nurses aides are available around the clock from Olsten Health Care.

"We have an obligation to provide persons who can meet clients' needs, so we have to make sure applicants are qualified. If they aren't fully skilled and are willing to be trained, we'll provide the assistance to qualify them," says Luneski.

Temporary services have also grown into other specialized areas, from accounting to technical industrial needs, aided by Olsten's Profiler, which custom matches temps to specific requirements.

This data system surveys the client company's characteristics and covers skill requirements, equipment, work hours, and other facets, giving Olsten a readily accessible description of the range of temporary help needs for that firm. Olsten is able to crossmatch the client's needs and the temp's application instantly.

Luneski is proud that Olsten has had the opportunity to contribute to and share in the spectacular growth of Bergen County's economy in the past 25 years. As a member of the business community, employing many thousands of Bergen County residents over this period, the entire commercial, educational, and cultural establishment has benefited from the infusion of additional payroll dollars.

At the same time, through Olsten, jobs have been available and employment furnished to those thousands of Bergen County residents whose life-styles precluded full-time permanent employment. Homemakers, teachers, students, senior citizens, and others have thus been accorded the opportunity to work when available and maintain their family responsibilities. The individual, the family, and the community all benefit.

JERSEY PRINTING
AND OFFICE SUPPLY CO., INC.

E. George Dabagian, president and founder of Jersey Printing and Office Supply Co., Inc.

One of Jersey Printing's multicolor presses.

Ask a group of business executives to name the most competitive business-to-business field and it is extremely likely that they will identify the printing profession. Within northern New Jersey there are hundreds of printers of varying size, quality, and ability—each seeking the business of the thousands of companies in the region.

In such an environment skill and quality are the expected norms. Success belongs to those possessing a special something extra—a commitment to service unsurpassed by its competitors. Jersey Printing and Office Supply Co., Inc., of Paterson has distinguished itself in that pantheon of North Jersey's leading printers for businesses and institutions.

Founded in 1952 by E. George Dabagian, an Armenian immigrant from Turkey, the company has grown from a single duplicator in a former tailor shop on Grand Street in Paterson to a formidable lithographer of the highest sophistication and quality. The diversity of the firm's output is mindboggling: from brochures to catalogs, annual reports to posters, prospectuses to technical manuals, and hundreds of other categories of fine color printing.

Jersey Printing serves a business clientele located throughout the northern reaches of New Jersey, into New York City, and even in the Midwest. Among its repeat customers are advertising agencies, nonprofit organizations, major medical facilities, and commercial banks and other financial institutions, as well as leading pharmaceutical companies and industrial concerns.

Having established the firm as a leader in the field, George Dabagian looks to the future and the role the company will play under his son, Richard G., vice-president of Jersey Printing, who is known throughout the company as "Rick." "We have set our course," relates the younger Dabagian, "on producing the finest-quality color printing for all applications by our customers."

Asked about the future economic viability of the North Jersey regional market, George Dabagian states emphatically that "this is conceivably the hottest area of the country... everything is at our doorstep."

And the customer is quick to take advantage of changing conditions in the marketplace, such as developing relationships with foreign-based companies operating in the state. "The decline of the dollar has opened up printing jobs that were done almost exclusively overseas," Dabagian notes.

Commitment to service and caring for its customers has been the hallmark of Jersey Printing and Office Supply Co., Inc., from its inception more than 35 years ago. That commitment is the inspiration for constantly upgrading equipment and fine tuning the training of personnel, practices that have brought the company plaudits and laurels from throughout the industry.

percent are enclosed malls. "There were 2,039 construction starts on shopping malls in the United States, alone, in 1986," he adds. Quoting figures from *Shopping Center World,* Pendley points to New Jersey's rank in average sales per square foot in malls: "At $152.48, New Jersey is third in the top ten for sales per square foot, after Alaska and Hawaii. When we look at the top ten states by total sales in shopping centers, New Jersey ranks ninth—neither Alaska nor Hawaii even make the top ten." In terms of large malls, New Jersey has its share. According to Pendley, the Garden State has a greater proportion of super-regional malls (750,000 square feet and up) than any other state in the Union. *Shopping Center World* credits New Jersey with 634 malls, with 22 (3.5 percent) falling into the super-regional category. New York State's percentage of super-regional malls is 2.3 percent; California has 2.4 percent; and Massachusetts only 1.1 percent. Pendley says, "The dominance of regional malls in New Jersey is far more substantial than in any other state. No one else even comes close. Clearly," he adds, "New Jersey has a great many good, comfortable, enclosed shopping malls."

THE MAJOR MALLS Northern New Jersey's spectacular array of malls contains something for everyone, from the average shopper to the most narrowly selective. Spread like a daisy chain throughout four northern counties, these big, beautiful malls make shopping convenient to residents wherever they are.

In Essex County, the Livingston Mall and The Mall at Short Hills predominate. Between the two, a shopper can savor merchandise that runs the gamut in variety from items in Sears to those in Bloomingdale's. While the Livingston Mall is anchored by Macy's, Hahne's, and Sears, The Mall at Short Hills has Bloomingdale's, B Altman, Bonwit Teller, and Abraham & Straus. Over 100 other top stores combine to create a sophisticated mélange of merchandising riches.

Morris County is no stranger to the regional mall, with Rockway Town Square mall sequestered neatly along Interstate Route 80. Anchored by Macy's, J.C. Penney's, Sears, and Hahne's, the mall offers over 200 smaller stores in an atmosphere created by modern sculpture, fountains, and plants.

Passaic County boasts two marvelous malls that literally abut each other: Willowbrook and West Belt. Willowbrook, which opened in 1969, has a staggering 175 stores and services in addition to Macy's, Stern Brothers, Steinbach, and Sears. West Belt Mall, its next-door neighbor, adds Fortunoff and J.C. Penney plus 23 stores and services to create a treasure trove for the shopper.

However, Bergen County's six malls, which exist within minutes of each other, make the county a diehard shopper's Valhalla. The state's oldest shopping mall, The Garden State Plaza, made history when it opened. Now newly refurbished and enclosed, it has lost none of its luster as one of the gems of the Northeast. Featuring 105 stores and services, the mall also offers Macy's, Hahne's, and J.C. Penney. Its quiet garden setting replete with trees and walkways creates an atmosphere where people can relax and shop serenely. The Bergen Mall adds Stern Brothers, Steinbach's, JJ Newberry, and 80 more stores and services only seven-eighths of a mile up the road from the Garden State Plaza. Of the major centers in Bergen, Paramus Park has the most stores—120 plus Abraham & Straus, Sears, and Fortunoff's. The Fashion Center is small but it has more than a touch of class, with Lord

Nationwide, Macy's customers are assured of the latest styles and top quality merchandise. Photo by Rich Zila

The atrium-like effect of the skylights and planters at the Woodbridge Center Shopping Mall creates a pleasant environment for shoppers. Photo by David Greenfield

Riverside Square Mall is located on Route 4 West in Hackensack. Photo by Carol Kitman

& Taylor and B Altman as its major tenants. And its 28 other stores live up to that standard. The Mall at IV is the newest and smallest addition to the Route 4 area, but its stores—Filene's Basement and Royal Silk—make it a unique expedition for any shopper.

The last, but in no way the least in reputation, is Riverside Square Mall, a multilevel mall also on Route 4 West in Hackensack. In this elegant mall, shoppers can choose from Bloomingdale's, Saks, Conran's, and 91 other stores and services. Elizabeth Napoli, general manager, describes the mall's unique image: "Riverside has carved out a niche in the marketplace which is all its own, and it is determined by our department stores: Bloomingdale's and Saks. We have similar de-

High standards of security are maintained at the Newport Centre mall.
Photo by Carol Kitman

mographics to the Short Hills Mall," she says, "but we have a much more fashion-forward customer. As a result, we are more trendy as opposed to a more conservative Ralph Lauren look. We are on the cutting edge of fashion," she adds, "faster to pick up a trend. Our customer is more willing to take risks earlier in the fashion season." Commenting on Riverside's success in such a mall-rich area, Napoli credits a realistic strategy: "We are competitive when we have to be; when everyone else is on sale, we're on sale. During special major gift periods, we promote ourselves as a destination point for the 'special' occasion." She also points out that Riverside is considered a base for items for the home, featuring Conran's, Bloomingdale's furniture, and other home stores such as the Pottery Barn, the Workbench, and more. But success, even in a great market like Bergen, is not assured. Napoli sums it up by saying,

It's a highly competitive place to be; there are constant changes in the market. You can't hope to be everything to everybody. If you don't do a good job, you will go out of business. These customers are savvy and especially sensitive to good sales and marketing techniques. As affluent

as they are, they are value conscious and they will shop at stores that have marketed to them and communicated with them.

Even with a heavy proliferation of malls so close together, some stores in the same chains are opening duplicate units in malls that are literally "next-door," or are opening stores on both Route 4 and Route 17 to take advantage of the flow of traffic in both directions.

A MALL WITH A VIEW One of northern New Jersey's newest shopping areas overlooks Manhattan's skyline and sits next to the mouth of the Holland Tunnel: Newport Centre, in Jersey City, which officially opened in October 1987. Two of the mall's four retail anchors, Sears and Stern's, are opened, but J.C. Penney's is not expected to open until 1989. The fourth will be announced later.

Perched on what has come to be known as northern New Jersey's "Gold Coast," the mall will have some of the nation's hottest retailers

Houlihan's Old Place is a fine restaurant refuge located within the Riverside Square Mall. There, shoppers can meet, eat, drink, and make merriment. Photo by Rich Zila

The Paramus Park Shopping Mall houses the largest number of stores of any Bergen County mall. Photo by Carol Kitman

Neighborhood storefronts, such as these located in Rutherford, remain as an alternative to the current shopping mall trend. Photo by Rich Zila

filling 165 spaces for specialty shops, including a 20,000-square-foot The Limited store on two levels. There will also be a Wallach's, Eddie Bauer, and Aggio (Webster Clothing's prototype of a trendy, upscale menswear shop). Aeropostale, which features trendy sportswear with an aviation theme, is Macy's new specialty store joining the others.

The Newport Centre mall is the centerpiece in the massive mixed-use development undertaken by Indianapolis-based Melvin Simon and Associates and the LeFrak Organization of New York. When it is complete in the mid-1990s, Newport Centre will encompass 9,000 residences, office space, two or three hotels, a marina, a cultural pier, an oceanographic museum/aquarium, parks, and more.

SHOPPING THE OUTLETS Northern New Jersey shoppers are not deprived when it comes to outlet shopping either. Cost-conscious consumers can find countless bargains at manufacturers outlets. Though most of these stores are less glamorous than those found in the giant suburban malls, the smart shopper can find tremendous values on high-quality merchandise.

The Meadowlands Parkway area in Secaucus is northern New Jersey's outlet paradise. With over 100 manufacturers' warehouse stores, the Secaucus Outlet Center offers the likes of Gucci, Calvin Klein, Liz Claiborne, Bally Shoes, and many more. Shoppers come from as far as Massachusetts, Connecticut, and Washington to take advantage of spectacular savings. According to Joan Hochmeyer, president of the Secaucus Outlet Center, "it is a unique and active area. I know people who fly in from Washington just for a day of great shopping."

DOWNTOWN AREAS Amid all the retail activity in northern New Jersey, its downtown areas flourish, unlike those in other areas of the country. Bergen County, in particular, retains the individuality and vitality of its downtown retail areas. This aspect of the area is corroborated by Elizabeth Napoli, who points out that "Every level of retail, from downtown areas to regional malls, is thriving and active." Napoli adds, "With all the malls in Bergen County, there are also very viable downtown areas. For example, Fort Lee is a wonderful place to shop. With its many transplanted New Yorkers strolling through interesting trendy shops, it takes on a delightful village atmosphere."

The urban retail marketplace provides sharp contrast to the shopping mall environment. Photo by Michael Spozarsky

The intersection of North Walnut and East Ridgewood Streets in the town of Ridgewood is a thriving commercial district despite the shopping mall boom. Photo by Rich Zila

COMMUNITY INVOLVEMENT   Most people in this country are cognizant of the role malls play in their daily lives. As a place where people congregate, malls have always been in a position to provide an arena for public service programs. Several projects of consequence have been initiated by the International Council of Shopping Centers in malls around the country, including northern New Jersey's.

In 1987 and 1988, ICSC has made battling the nation's alcohol and drug problem a priority. "Last year the program, known as 'Kids Know,' was in place in 2,200 shopping centers nationwide," public relations director Donald Pendley said. "This year we hope to exceed that number." Shopping centers will sponsor information exhibits and volunteers who will dispense literature on the topic. "The thrust of its message," says Pendley, "is that kids know more about drug problems than we give them credit for, and they need to learn to build kid-to-kid networks. When a child is offered drinks or drugs," he said, "parents are never around, so they need the support to say no from each other."

Many malls in the north conduct other programs that benefit the members of each community. In particular, several malls open their doors early so that senior citizens can walk safely for exercise before the stores open for the general public. There are walking clubs, and areas are set aside where seniors can change.

years. I can't believe New Jersey will be immune to it. If there is going to be expansion along entertainment and social kinds of functions, which I think is inevitable, New Jersey is more subject to it because the New York population won't be able to do that sort of thing.

Pendley calls New Jersey a bellwether state for what is happening in the industry. "If you look at the industry over the past three decades and where it is going, New Jersey would serve as an indicator. Southern California," he says, "and Florida are also areas to watch; but, interestingly, New Jersey is the only one of the three not experiencing population growth, and still it serves as an accurate gauge for the industry."

Over the years, malls have become a staple in our daily lives. They have long been touted as the new "town square" or community center. According to ICSC, in any given month 94 percent of the United States population 18 years or older visits a center at least once, on the average at least twice. The malls have also created nearly $7 million in permanent jobs for people, or about 7 percent of the United States' non-agricultural employment. Needless to say, malls are part of the fabric of our society. ICSC's Pendley extrapolates that "if you look at all the institutions within a community, I don't think that any serves the economic and the social under one roof the way the shopping center does. If there is a place away from the home," he says, "where family life can happen, it is at the mall." In fact, malls have even spawned their first genuine rock star. In a promotional coup, the rock star Tiffany was created when she decided to go where the teens were—the malls.

The idea of "the mall" has helped to conjure many a metaphor and inspire reams of lofty prose. Few things have burrowed into the American psyche and lifestyle with such rapidity (although the first mall was created in the twenties in Kansas City, malls have only been common since 1957). Yet writers continue to rhapsodize about them, and to ponder their social significance. In her essay entitled "On The Mall" from her collection entitled *The White Album*, Joan Didion asserts that "They float on the landscape like pyramids to the boom years …" Northern New Jersey's many malls, its healthy retail climate, and its new construction activity offers tangible evidence that the region is still in the throes of its "boom" years.

Stores at the Riverside Square Mall cater to individuals with up-to-the-minute taste for clothing. Photo by Rich Zila

WHAT THE FUTURE MAY HOLD   The growth of the shopping mall as a social/entertainment center is an idea whose time may be at hand. With the opening of West Edmondton Mall in Alberta, Canada, the shopping mall may take on a new twist here in the United States too. According to Pendley, "We are seeing a trend now between melding the shopping center and the theme park." The mall in Alberta combines an amusement park, shopping center, hotels, a man-made lake (complete with a submarine and a model of one of Columbus' ships), 828 stores, a zoo, and a skating rink all under one roof on six million square feet of land. "They are building one, only bigger, in Minnesota," says Pendley.

If it catches on here, you will find it happening in the next five to eight

A tremendous rate of growth is apparent in the New Jersey banking industry. Photo by A. Satterwhite

New Jersey's reputation as a financial innovator and leader goes back to our nation's beginnings, when Mark Newbie became, in effect, the first American banker. Newbie, a Quaker from London, opened and operated the first bank in the new world. The "bank" was actually a room in his home near Camden, and his capital consisted of 300 acres of land offered to secure the currency he used: copper coins from Ireland known as St. Patrick's Pence. Newbie's efforts constituted the first authorized issue of currency in the colonies when his idea was accepted by the General Assembly at Burlington in 1682.

The state, and the nation, has come a long way since the time when settlers found it convenient to adopt the use of Newbie's Irish coins.

CHAPTER FIVE

With interstate banking signed into law in 1986, New Jersey's financial institutions have found it easy to keep pace with a rapidly changing marketplace.

A HEALTHY BANKING ENVIRONMENT   The state's vibrant and diversified economy has assured prosperity for its many banks. With businesses continuing to flock to the state, the economy is boom-

Growth And Prosperity In Banking

United Jersey Banks of Princeton is the third largest banking corpora- tion in New Jersey. Photo by Bob Krist. Courtesy, Jim Adamczyk, United Jersey Banks

ing. But New Jersey's prosperity is no longer news; it has continued unabated for seven years with no signs of deterioration, and the state continues to outperform the nation as a whole. The state occupies sec- ond place nationally in personal income, behind Connecticut, with a 44 percent increase during the last five years alone. The transportation infrastructure is the best on the East Coast, with nearly 80 percent of the combined New York/New Jersey port volume of foreign trade han- dled in the Garden State. New Jersey outpaces the nation in retail sales, personal income, non-residential construction, and employment.

The state's robust economy has created 521,000 new jobs since 1979. In fact, employment has increased in the state at a rate of nearly 50 percent greater than the rest of the nation. Unemployment in New Jersey, now at approximately 3 percent, is the second lowest of all 50 states.

All of these factors contribute to making New Jersey the outstand- ing success story that it is. That healthy economic environment, in turn, provided the essential ingredient for financial institutions to thrive. Commenting in the *Star Ledger*'s Outlook on the symbiotic relationship between banking and New Jersey's economy, Robert Van Buren, chairman of Midlantic Corporation and the New Jersey Bank- ing Association, said:

New Jersey's commercial banking industry has both supported, and benefited from, our state's economic vitality. As a result, the industry is in the strongest position in history. The earnings and key performance ratios are the levels appropriate to support future growth and develop- ment of our communities, which in turn will contribute to further expansion.

Mary Little Parell, chief executive of the New Jersey Department of Banking, points out that banks in the state, unlike banks in many other parts of the country, have not reported diminished earnings. In- stead, on June 30, 1987, New Jersey's state banks reported a 22 percent increase in earnings over the first half of 1986. "Likewise," Parell said, "New Jersey's savings banks disclosed a 15 percent boost in earnings during the same period." Clearly, the strong state economy and the state banks' non-involvement with foreign loans have nurtured the growth and prosperity it now enjoys.

As one of the strongest areas in the country for both commercial and consumer banking, New Jersey is the ninth-largest banking market in the United States. The enviable position of the state's financial insti- tutions have made them an enticing target for mergers and acquisitions. In 1987, New Jersey banks saw interstate banking explode on the scene, spurred by the continuing profitability of its banks and the econ- omy within which they operate. The regional phase of interstate bank-

This United Jersey Banks branch office is located in Rutherford. Photo by Rich Zila

ing, signed into law in April 1986, is the precursor of the national phase of the same law, which was activated at the start of 1988. Sixteen states became eligible as of January 1 for reciprocal interstate banking with New Jersey, subject to regulatory approvals. With the new law, New Jersey banks and their stockholders now can entertain offers from the money center banks in New York. Concurrent with all of this activity, the state experienced a spurt in the chartering of new community banks, which have been created to fill perceived vacuums at the local level. All in all, after a year of tremendous growth and change, New Jersey's banks have never been in better shape.

Horizon Bancorp headquarters located at 225 South Street in Morristown is also the location for the main branch of Horizon Bank, one of its seven subsidiaries. Courtesy, Horizon Bancorp

William J. Shepard is the president and chief executive officer of the Horizon Bancorp. Courtesy, Horizon Bancorp

ENTERING A NEW ERA As of this writing, five of New Jersey's largest commercial banks have been involved in pending or completed mergers that cross the state's two famous bordering rivers—the Hudson and the Delaware. They include the $17-billion-asset Midlantic Corporation of Edison; the $29-billion First Fidelity Bancorporation of Newark; United Jersey Banks of Princeton with over $10 billion in assets; the $4.4-billion First Jersey National Corporation of Jersey City; and Horizon Bancorp of Morristown with $4 billion in assets. Three of the five banks involved in these transactions have their roots in northern New Jersey. (All asset figures listed in this chapter were compiled in January 1988.)

First Fidelity Bancorporation of Newark agreed to acquire Fidelcor Inc. of Philadelphia on July 31, 1987, in a stock exchange valued at $1.34 billion. The transaction created a $27-billion "super-regional" bank. The merger will be the highest-valued ever in banking history. First Jersey National Corporation was chosen for acquisition by England's largest—the $140-billion-asset National Westminster (NatWest) Bank Group of London. This is the state's first foreign bank takeover; it will unite First Jersey with NatWest's New York subsidiary. The merger was approved in December 1987 by the Federal Reserve; it became effective February 1, 1988. The transaction, valued at $820 million, represented a premium of 50 percent over the market price at the time. Horizon of Morristown has agreed to an acquisition by Chemical Banking Corporation of New York in what would be the first transaction of its kind between bank holding companies in two states.

Both banks have decided on a January 1, 1989, starting date, even though interstate banking became legal in New Jersey this year. The cost to Chemical is estimated to be more than $600 million.

Midlantic Banks Inc. of Edison completed a merger on February 1, 1987, with Continental Bancorp of Philadelphia in a stock transaction valued at $680 million. The combined assets equal $17.2 billion, which made Midlantic the state's largest bank at the time of the transaction. Midlantic Corporation was created to take ownership of the two banks. United Jersey Banks of Princeton, the third-largest bank in the state, acquired First Valley Corporation of Bethlehem, Pennsylvania, for an estimated $265 million in stock. The transaction will create a regional bank with over $10 billion in assets.

According to Mary Little Parell, "banks have gained flexibility during their passage into a new regulatory and competitive environment during this decade. Thus, their plans," she continued, "reflect an awareness and an ability to cope with the fluctuating conditions in the national economy even while participating in positive forces at work in the New Jersey economy."

National Community Bank, headquartered in Maywood, is one of the best performing banks in the state, and it remains an anomaly. While it adheres to its basic philosophy to remain an independent, homegrown bank, its series of community banks strewn throughout the state have combined assets of over $3.2 billion.

Below: First Fidelity Bancorporation headquartered at 550 Broad Street in Newark. Courtesy, First Fidelity Bank

Right: Hackensack is the largest member bank of the United Jersey Banks corporation. Courtesy, United Jersey Banks

The headquarters of the National Community Bank is located in Maywood. Photo by Rich Zila

Robert M. Kossick, National Community Bank's president and chief executive officer, said in a recent interview: "I think many people are distraught with the mergers. There is going to be a great deal of disruption and turmoil, with more unhappy than happy customers." Kossick continued, "We stay close to the customer ...Many people believe that a bank's product is its money. It isn't; it is its people. That is what a bank has to offer." Describing money as an undifferentiated product, Kossick said, "People can go to any bank and get it. We believe customers go to banks for the people, the service, and the attitude that exists." Kossick said that mergers have tended to make banks colder, less available, and more apt to rely on machines and technology rather than their greatest natural resource—their people.

CITIZENS FIRST NATIONAL BANK

In 1934 the bank building located on the corner of Prospect Street and East Ridgewood Avenue became Citizens First's headquarters and main office.

Northern New Jersey has undergone dramatic changes since a small group of business leaders met in 1899 in the village of Ridgewood to lay the foundation of what would become Citizens First National Bank: the first bank established in northwest Bergen County.

From this auspicious beginning the bank has grown in size and stature, as well as in the sophistication of its services, to match the unparalleled development of northern New Jersey.

Since opening for business in an abandoned grocery store on July 24, 1899, Citizens First National Bank has broadened its service base through acquisitions and by opening new branches. Today the bank operates a network of approximately 50 offices in key growth areas of the Garden State—with the predominance of its operations still in the northernmost portions of the state and with its main office still located in Ridgewood.

As the bank and the region approach the end of the twentieth century, Citizens First operates from a modern executive center in Glen Rock, which draws the wide-ranging branch network together with the bank's 10 operating divisions.

The expansion of facilities and the planned growth of Citizens First is based on the original precepts of the early

founders: responsive service to the people and businesses within the bank's marketing areas. Northern New Jersey is easily one of the premier growth areas of the nation. This reality, coupled with the state's increased economic and population growth in Ocean County,

The administrative headquarters of Citizens First National Bank is located at 208 Harristown Road in Glen Rock, New Jersey. Built in 1980, this visual symbol of growth and progress enabled the bank to consolidate its divisions and departments under one roof and provides for anticipated growth requirements for the future.

plus the steady expansion in Morris County and the revitalization of Hudson and Passaic counties, helped in determining the bank's projection of planned growth management. Strict attention to these initial principles is one of the reasons that Citizens First has succeeded as a provider of financial services for its commercial and retail customers, as well as a profitable enterprise for its investors.

Although the activities of the 10 divisions of Citizens First may be diverse, their central purpose is the same—expanding the bank's services, assets, and earnings for the benefit of all its customers, especially the small and medium-size businesses identified as the primary growth market for the bank.

Situated alongside administrative headquarters, the bank's Operations Center houses the Operations Division, which includes the many departments of this important bank function—processing and recording up to 300,000 separate paper items and processing up to 75,000 on-line transactions on a daily basis.

The Lending Division consists of highly specialized commercial, consumer, and international credit operations, and represents Citizens First's

Over the years Citizens First National Bank has sponsored the annual March of Dimes Sports Awards Banquet. Richard G. Kelley, chairman of the board (left), addresses the guests and honorees. He is assisted by Marv Albert, sportscaster, and Rodney T. Verblaauw, president.

largest segment of earning assets. Under the umbrella of this division's services are every credit-based transaction, from personal auto loans and home mortgages to major business construction financing, as well as international banking services that enable local businesses to pursue import and export opportunities in foreign lands.

Other divisions providing direct customer services include Corporate Banking, Investments, Trust, Correspondent Banking, and Branch Administration. They furnish commercial and individual customers with specialized services and opportunities in such areas as businesses, nonloan financial arrangements, personal and corporate financial planning, interbank transactions, and servicing and retail consumer services.

Some examples of this service are the Corporate Banking Division, which, together with the bank's network of local branch offices, can offer businesses such services as escrow accounts, rental security, account reconciliation, gov-

ernment banking, short- and long-term loans, lines of credit, real estate construction, and EDA loans. Personalized Services covers checking, savings, MasterCard, safety deposit, and Treasurer automatic teller machines. These services are enhanced by the bankwide cross-selling programs that provide existing customers with the opportunity to learn about additional and helpful financial services.

Behind these direct service divisions are the bank's support structures, such as Administrative Services, Accounting, and Personnel, which are vitally important to the smooth and efficient delivery of services to customers.

Citizens First National Bank is the wholly owned subsidiary of Citizens First

Bancorp, Inc. This corporate structure has grown to exceed $2.5 billion in assets and an employee roster of more than 1,000 professionals serving in diverse banking positions.

As in the past, each office of the bank has involved itself in the communities it serves, not only in sponsorship of local charities, sporting events for youngsters and adults, hospitals, scouting, red cross, and many other philanthropies, but also through substantial employee involvement in community service.

If the past is prologue to the future, the bank can look forward to an exceptionally bright future. Citizens First National Bank pledges not to rest on past laurels. The enormous shift from a primarily consumer orientation to a strongly commercial direction over the past decade testifies to the will of the bank's management and staff to succeed in turning the challenges of the future into opportunities for present customers, employees, and investors.

Kossick points to an even more serious problem involving the trend toward mergers:

I am against the interstate mergers of banks. By allowing full interstate banking in New Jersey, we will soon turn over control of our financial destiny to out-of-state banks. I don't want to see New Jersey borrowers having to go to New York or Philadelphia for loan approvals. New Jersey is an entity unto itself, not a place to pass through on the way to those cities.

Kossick underscores New Jersey's wealth and financial importance by calling it "the crown jewel of this eastern quadrant of the United States."

Kossick points to two major factors that have been responsible for the state's advantageous position in the marketplace. "Firstly, America's strength is becoming increasingly concentrated on its coasts," he explained, "because we have become a nation of importers and exporters tied into service-based industries." Kossick went on to say that the East Coast is the stronger of the two because of its link to Europe.

Outlining what he believes to be a second factor contributing to New Jersey's economy, Kossick calls the state the "administrative and technical back office" for New York. "New York is a money center," he said,

and between now and the year 2000, it will evolve into the world's pre-eminent financial center. As that happens, it will drive all the infrastructure in administrative and data processing into New Jersey, where, in most cases, a company can get a 25 percent to 30 percent improvement on a bottom-line basis in terms of operating expenses, a better work force, and more space.

National Community Bank President and Chief Executive Officer Robert M. Kossick maintains that banking is a service-oriented industry and that the personal touch is essential. Courtesy, National Community Bank

Kossick uses several examples of this trend as evidence of the business movement into New Jersey: "The examples go on and on," he said. "Plaza Technologies, which is the back office for Solomon Brothers, moved to Lyndhurst near Giant's Stadium; Alliance Capital moved into Secaucus; Paine Webber moved into Lincoln Harbor in Weehawken; and Prudential Bache moved to Edison." Kossick also

National Community Bank was originally chartered as the Rutherford National Bank. Its purpose was to service employees of the Becton-Dickinson Corporation whose branch facility is seen here. Photo by Rich Zila

Broadway Bank & Trust Company President Peter M. Kolben, standing, and Chairman of the Board Martin Sukenick, seated, have shifted the bank's interests from traditional lending to equity participation. Courtesy, Broadway Bank & Trust Company

notes that there is talk of Citicorp and Chase moving back-office operations across the Hudson River. "That will be the continuing strength of New Jersey," he said. "We will be the administrative and technological back office of the 'world's' financial center."

National Community Bank "played a major role in developing the Meadowlands when it was merely a dream," Kossick said. Calling the Meadowlands the greatest commercial real estate success story in New Jersey history, he cited the singular efforts of Fairleigh S. Dickinson and Hartz Mountain for their foresight. "It is easy to take pride in New Jersey now, but those original major movers of the project had to have an awful lot of belief that something could happen in what was then a swamp area." The last 15 years have borne out that belief, as the Meadowlands area has metamorphosed into the finest sports complex in America.

Broadway Bank in Paterson is another bank that has found its niche within a changing marketplace. Peter M. Kolben, president of Broadway Bank, commented on the effect of deregulation on the bank: "I think there is a real place in the market for banks like ours. All our competitors have been bought up, and it is probably the best thing that has ever happened to us." He went on to say that "we find it easier to pick up the business of people who still want the 'hometown' touch. They know they have access to the president and the assurance of continuity in the people taking care of their company, so our problem has been 'containing' growth, not 'sustaining' it." After 65 years in Paterson, Broadway Bank intends to stay independent. "We want to

GRANT THORNTON, ACCOUNTANTS AND MANAGEMENT CONSULTANTS

In the conference room with Grant Thornton. From left are Brian Downey, audit manager; Bill Haggerty, audit partner; John O'Leary, audit department head; Nancy Petrovich, office manager; and Mike Signor, audit manager.

A partner discussion in the office of the partner-in-charge. Pictured here (from left) are Howard Cohen, partner-in-charge; John O'Leary, audit department head; and Bill Murphy, audit manager.

Northern New Jersey's economy is on the move and brimming over with companies of many descriptions. While many of these corporate residents in the area are large, well-established, mature firms, many more are emerging and growing companies in America's most dynamic business sector, the middle market.

Regardless of a business' size, longevity, or field of endeavor, its management requires seasoned professionals to help plan the company's growth in size, profitability, and scope. Grant Thornton,

Accountants and Management Consultants, provides that level of service for clients locally, nationwide, and even worldwide.

The kind of professional assistance offered clients depends on several factors: the nature of their businesses; the size, characteristics, and internal resources of their organizations; and the specific financial issues faced by management. Following an integrated approach to each client's situation, the Grant Thornton team provides sophisticated, timely, and responsive business advice.

Grant Thornton goes beyond the traditional tax, audit, and consulting services to provide clients with the full spectrum of business assistance. As the firm's relationship grows with clients, it

can better match them with the opportunities they seek. Whether through the firm's capital markets group, introducing clients to merger or acquisition candidates, assisting them in bank and equity financing, or helping take their company public, Grant Thornton can evaluate the opportunities and help structure the transactions to assure maximum business opportunities and minimum tax liabilities.

The firm is organized along three separate, but integrated, disciplines: accounting and auditing, tax planning, consulting and compliance services, and management consulting services.

As auditors, the firm becomes steeped in every aspect of clients' businesses. With this in-depth understanding of client goals, the firm can make recommendations designed to increase profitability and increase overall efficiency. In supporting and supplementing clients' internal resources, Grant Thornton is prepared to analyze the financial, organizational, and operational impact of mergers, acquisitions, and divestitures; assist in securing financing or meeting financing requirements; review and analyze information-gathering and reporting systems; assist in the development of capital and expense budgets to help monitor and control costs; develop or improve inventory control systems; and consolidate accounting systems.

Specialists assist clients in dealing with the complexities of the tax environments in the United States and abroad. Grant Thornton provides a comprehensive range of tax-related services and activities. These include advising of tax benefits of business acquisitions, combinations, and mergers; developing current and deferred executive compensation programs; consulting on the tax feasibility of real estate and other business investments; individual and corporate tax planning; advising clients on the tax aspects of international

A tax planning discussion with (from left) Clarence Kehoe, tax manager; John Simpson, tax partner; and Dan Moore, tax senior.

business; providing assistance in complying with the tax requirements of qualified pension and profit-sharing programs; planning, preparation, and review of federal, state, and local tax returns; and representation before all levels of the Internal Revenue Service or state tax authorities.

The firm's management consulting services apply to virtually every aspect of a client's business, from the operations and financial areas to organizational development, information systems, and human resources management. Grant Thornton consultants assist clients in reducing costs, increasing efficiency, and solving a variety of business problems wherever they occur—in information systems, materials management and distribution, organization and staffing, or cash management.

Grant Thornton is an international firm with offices in 65 major commercial centers throughout the United States, including its northern New Jersey office in the midst of the Prudential Business Campus in Parsippany, in addition to offices located in 55 other countries through Grant Thornton International. From its Parsippany location Grant Thornton's teams of professionals can provide rapid response to clients in virtually every corner of the northern Jersey market.

Nationwide, the firm's 3,500 staff members, including nearly 1,000 partners and managers, provide comprehensive business services in accounting and auditing, as well as tax and management consulting. Of this number, 40 professionals are employed at the New Jersey

facility, where they furnish full services locally and provide a conduit to the national resources of the firm. This large management base helps the firm maintain a close, personal relationship with each client, while its size and scope supply depth of technical experience across a full range of services.

Grant Thornton's practice represents virtually all types of enterprises, both publicly and privately held. Clients include manufacturers, wholesalers, retailers, contractors, real estate developers, financial institutions, utilities and energy developers, health care agencies, hotels and resorts, governmental agencies, and not-for-profit organizations.

The firm's entrepreneurial approach—solving problems across

Rich Dannenbaum, audit senior, utilizes the newest in computer audit techniques.

traditional lines with hands-on involvement by top management—fits the style of its enterprising clients. This harmony invites creative solutions, practical recommendations, and prompt implementation.

Each client is served on a team basis, providing access to the various experts within the firm who can be called on to deal with specific situations. Further, the client benefits from Grant Thornton's strong ties within the banking, legal, and business communities in fostering business relationships to help clients grow.

Full resources can be brought to bear on the particular needs of the client as necessary, based on six basic tenets of the firm: national presence, technical excellence, full range of services, skilled professionals serving as business advisers, experience with a diversified client base that helps Grant Thornton anticipate needs, and quality service that adds value to the client's operations.

Headquartered in Chicago, Grant Thornton, Accountants and Management Consultants, is positioned to meet the needs of today while assisting clients to take advantage of the opportunities of tomorrow. Grant Thornton—"We help businesses grow."

MIDLANTIC NATIONAL BANK/NORTH

Midlantic/North's headquarters building rises above Interstate 80 in West Paterson, linking the bank's offices in Passaic, Bergen, Essex, Hudson, and Morris counties.

From the beginnings of its first predecessor, the Paterson Savings Institution, founded nearly 120 years ago, Midlantic National Bank/North has been in the vanguard of progressive commercial banks, nurturing the economic development of northern New Jersey for individuals and businesses.

Today the bank has more than 70 offices, serving communities throughout the northern tier of the state with the most modern financial services available and backed by the resources of Midlantic Corporation, one of the leading regional financial services companies in the nation.

It is the tradition of community identification and service that has always set Midlantic/North apart from its competitors. Despite growth that has taken the bank from its first city origin to dozens of large and small municipalities in Bergen, Passaic, Essex, Morris, and Hudson counties, the bank is found at the forefront of leading community efforts to support important local causes, from youth services to the concerns of the elderly. Virtually every project benefiting the economic and social conditions of its various service locations has a banker from Midlantic/North involved in some meaningful way.

It is as an innovative commercial bank that Midlantic/North has its most profound impact. Whether serving an international conglomerate, local contractor or manufacturer, community institution, neighborhood store, or individual consumer, the bank's responsiveness to its customers' needs is of primary importance. And the bank has changed the scope of its services to meet the constantly changing needs for those services.

As the economic environment in northern New Jersey grew, so did Midlantic/North, both in size and in the diversity of its services. The bank introduced such innovations as the first automatic teller machine system in North Jersey, the first international banking department to help local businesses expand into foreign trade, consumer credit services, and other trendsetters that are now considered standard for banks.

In recent years the financial services industry has seen many changes. Through its membership in one of the most dynamic banking organizations in the United States, Midlantic/North has succeeded in keeping pace with these changes and, indeed, turning them into opportunities for businesses throughout the region.

Midlantic National Bank/North will continue to serve the expanding needs of northern New Jersey's growing economy, thereby ensuring continued prosperity for local communities, businesses, and the people living there.

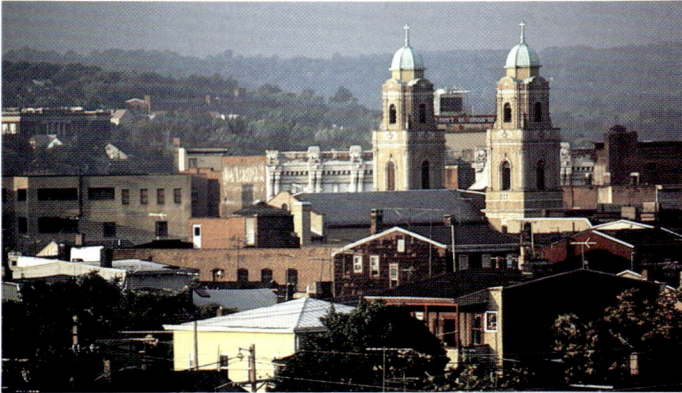

A distinctive feature of the downtown Paterson skyline are the steeples of St. Michael's Church. Photo by Michael Spozarsky

continue to serve our customers and their unique needs," Kolben said. "There are many companies that need our services."

Diversification has been the backbone of northern New Jersey's longstanding business boom, and now it promises to do the same for Paterson's economy. "Paterson has learned the devastating effects of being a one-industry town," Kolben said. "Now the old mills are occupied by a multitude of different companies, and we are surviving much better." As an example of the direction in which Paterson is embarking, Kolben points to the development of "Superblock," a mixed-use project already underway. "Paterson's potential for redevelopment is looking better and better," he said. "Based on plans for the highway to come right into the city, developers are willing to build a 1-million-square-foot office, parking, retail, and residential real estate project. The federal government plans a 50,000-square-foot administrative building, and the county is also planning to consolidate all its facilities here."

The ultramodern home of Broadway Bank, the 14-story Hamilton Plaza office building, lies in the center of Superblock. As a relatively new addition to Paterson's skyline, Hamilton Plaza provides the first piece to the project, which is in progress. The modern complex is juxtaposed with Paterson's historic district, which was declared a national landmark in 1976. As the first planned industrial city, Paterson's roots extend to the time when Alexander Hamilton, President Washington's secretary of the treasury, foresaw the potential industrial strength offered by the city's 77-foot-high, 280-foot-wide waterfall. "The city is unique because the old and new coexist, which gives it charm," Kolben said. "Pierre L'Enfant, the architect who later designed Washington, also laid out Paterson. The story goes that he learned from his mistakes. This city embraces a magnificent history."

In discussing the future of northern New Jersey's newly revitalized cities, Kolben notes the numerous advantages they possess:

Cities aren't going to disappear. Employment opportunities must be provided for everyone, and now that is happening. There are companies that chose nice country locations in the 70s, only to be thwarted by the gas crunch. When their employees could not drive to work, they realized the value of the city's ready-made infrastructure and labor force and moved back. The sewers, electric, water, buses, and railroads are all here waiting to serve people.

Commenting on Paterson's future, Kolben said that he was optimistic. "We want to see Paterson do well, and we are in a position to influence that. Banks aren't removed from the communities in which they reside. We can't do better than the community is doing. If there are no economic returns from the city, there is no return to us." Paterson's relatively small size is also an advantage; "we have 140,000 to 150,000 people," said Kolben. "Go three miles in any direction and you will find the beautiful, growing suburbs of Bergen County. Even Manhattan's midtown is only a 19-minute train ride away."

Broadway Bank's role involves servicing "a particular segment and its needs, and we do it well," Kolben said. "Most of the companies here are not multinational or national, but family-owned or small publicly held companies, and we have the resources to take care of their problems." Since deregulation, Broadway Bank has involved itself in diversified activities— real estate development in particular. "We are not new to this," Kolben said. "The bank has been in the forefront in residential development since the forties, fifties, and sixties. Many of the homes in Elmwood Park and Clifton, built for servicemen, were financed by our bank." Now the bank is involved as an equity partner in real estate development. "We do joint ventures with developers," Kolben said. "We see the broadening of our base from traditional lending to equity participation as a very important part of our future. We have more resources than the average developer, so we can be very helpful."

In assessing New Jersey's banks, Kolben gives them high marks. "Banks in New Jersey have more than kept up with the burgeoning growth in the state. I don't know of any companies that can't get the support they need to maintain their growth and business prosperity."

This Paterson renovation is a result of local banking interests in small business development. Photo by Bob Krist

New Jersey residents have access to both New York, and New Jersey, television and radio stations. Photo by J. Brignolo

Before the magic of television transformed the world into what Marshall McLuhan has called a "Global Village," newspapers were our major source of information. They made it possible to know what was occurring in the farthest reaches of the world. New Jersey's 26 daily newspapers were no exception. Over the years they performed their tasks well in a market that included the giant metropolitan dailies in New York and Philadelphia. In fact, it was the state's strategic placement between these two major cities during the American Revolution that prompted the creation of its first newspapers.

The history of the "Fourth Estate" in New Jersey begins on December 5, 1777, when Isaac Collins, with the support of the Republican legislature, began printing the *New Jersey Gazette* at Burlington. With the assistance of George Washington and Alexander Hamilton, Shepard Kollack started the *New-Jersey Journal,* known today as the *Daily Journal,* at Chatham on February 6, 1779. According to William Levin in his book *A Story Of New Jersey Journalism,* "Both papers, one in south Jersey and one in north Jersey, were created

CHAPTER
SIX

Media: Getting The Message Across

The main plant of **The Record** *is located here in this expansive building in Hackensack. The corporation does, however, maintain offices in Passaic, Trenton, and the National Press Building in Washington, D.C. Courtesy, The Record*

to champion the Revolutionary cause."

Today, New Jersey's location between New York and Philadelphia places it in one of the most competitive markets in the world. Five of the twelve largest newspapers in America are published and circulated within this area. With the *New York Times* and the *Philadelphia Inquirer* close by, the job of providing alternatives to these mammoth metropolitan papers is a tough one. However, New Jersey has fashioned a journalistic tradition that serves its market well. In his foreword to David Sachsman's and Warren Sloat's *The Press And The Suburbs, The Daily Newspapers of New Jersey,* George Sternlieb, director of the Center for Urban Policy Research at Rutgers, comments on the evolution of the state's unique journalistic approach: "In an intense effort to keep pace with the changing location of their readers—and most particularly with the upscale consumers—the shift to the suburbs has been marked by changes in news coverage, advertising, and promo-

tion." Sternlieb maintains that while this "suburbanization" is happening throughout the country, it is

epitomized in the most suburban of states—New Jersey. With perhaps the greatest saturation of television alternatives in the United States, with major cities of relatively small scale, New Jersey is subject to all of the competitive pressures. It thus provides a true test of the capacity of newspapers not only to survive—but to do so while maintaining credibility and a public function.

NEWSPAPERS IN THE NORTH   One of northern New Jersey's largest newspapers, *The Record* in Hackensack, embodies the highest standards of a good suburban daily newspaper. Sachsman and Sloat cite the results of a 1974 study conducted by the Center for Analysis of Public Issues in Princeton, in which New Jersey reporters and editors ranked the best in the state: "The journalists rated *The Record* the best all-around newspaper in New Jersey." Always garnering high marks for its excellent writing, Sachsman and Sloat call it the "best-written newspaper in New Jersey." *The Record* today continues to win accolades. *Time Magazine,* in its decennial rating of the best papers in the country, singled out *The Record* as a quality suburban newspaper. With a daily circulation of more than 156,000 and 220,000 on Sunday, *The*

Malcolm A. Borg is the current chairman of the board at **The Record.** *The Borg family has controlled* **The Record** *since 1930. Courtesy, The Record*

Record serves five counties, four of which are in northern New Jersey: Bergen, Passaic, Morris, and Hudson. It also serves neighboring Rockland County in New York. In addition to its main plant in Hackensack, *The Record* maintains offices at its Passaic/Morris headquarters, and has news bureaus at the statehouse in Trenton and the National Press Building in Washington, D.C.

With *The Record*'s long roots firmly implanted in Bergen County soil, it is no wonder that its knowledge of northern New Jersey is extensive. In its 93-year history, it has grown with the northern communities that it has served so well. But when it began on June 5, 1895, it was little more than a "small town" paper housed in a Hackensack storefront, next to a laundry. By 1901, it had moved to 131 Main Street, and finally, in 1951, it came to occupy its present location at 150 River Street. Over the years, its improvements and expansions have been impressive. As a result, *The Record* has become known for its state-of-the-art production technology and unique color printing capability. *The Record*'s high-speed, custom-designed offset presses make it possible to print four-color photos and graphics on 16 pages and spot color on another 48. *The Record* has the distinction of being one of the first newspapers in the nation to use computer-generated graphics.

Growing in step with the area in which it resides, *The Record*

serves one of the largest concentrations of retail business in the nation, and it is a leader in newspaper advertising linage nationwide. The paper currently leads the nation's Sunday newspapers in color advertising linage, and it is among the top 10 daily and Sunday combination newspapers in total advertising linage.

Over the years, *The Record* has been a consistent leader among the state's newspapers. In their examination of New Jersey's newspapers, Sachsman and Sloat explain its strong points by saying that its "in-depth interpretive and investigative pieces win journalism awards, and are considered by many the hallmark of quality journalism. This is *The Record*'s strength, and journalists across the state value it highly."

The Record has been under the direction of the Borg family since 1930, when Wall Street financier John Borg acquired control. Malcolm Borg, John Borg's grandson, is the current chairman of the board.

The Record's involvement with its community has taken many forms over the years. However, recently it has found a means by which to honor local athletes while fostering the qualities by which youngsters become vital participants in their communities. Through its Al Del Greco Award, *The Record* rewards high school students on the basis of leadership, sportsmanship, scholarship, and citizenship. Each week throughout the year, athletes are singled out for praise in a weekly

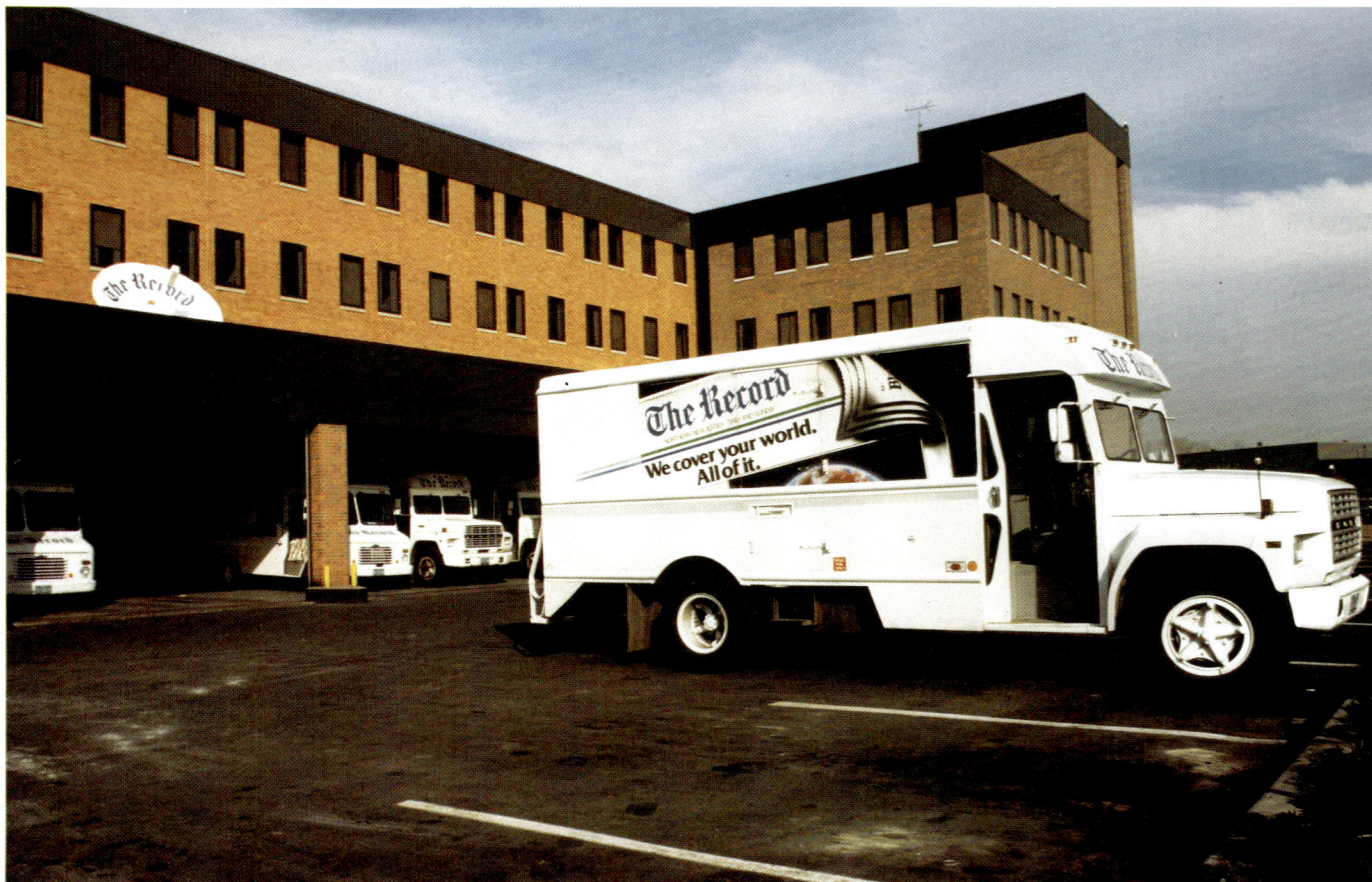

The Record *is the number one news-paper in Bergen County. Photo by Carol Kitman*

newspaper column, and are then honored at the annual Athlete of the Week awards banquet. The award itself is named for the late Al Del Greco, who was a sports editor at *The Record*. It was his belief that the true sports "stars" were not always the high-scorers or headline-makers; that philosophy is the basis by which students are chosen as recipients. In the past, Del Greco award winners have included handicapped athletes, outstanding team managers, and even a cheerleader who prevented an angry crowd of fans from turning into a mob. Each winner is presented with a replica of his or her portrait (which accompanies the sports column) rendered by Charlie McGill. McGill is *The Record*'s sports illustrator.

The *Star Ledger* is one of the leading newspapers in the Newhouse chain, which owns newspapers across the United States. It is the largest in the state, and the eighteenth-largest daily in the country. In terms of circulation, one out of every four daily newspapers and one out of every three Sunday papers published in New Jersey is the *Star Ledger*. The paper's excellent and thorough statehouse coverage makes it the paper of choice for those who want to know what is happening in New Jersey politics. This New Jersey giant's sphere of influence reaches from Essex County through Union, Morris, Middlesex, Somerset, Monmouth, and even Sussex to some extent. And its expansion continues unabated. The *Star-Ledger* has reporters in bureaus throughout most of the northern and central parts of the state. Its

sports coverage of schools and colleges is also its forte. In a series of articles about New Jersey's leading daily newspapers, Sachsman and Sloat admit that "the *Star-Ledger* is still the newspaper that accompanies the morning cup of coffee in homes, offices, and diners throughout New Jersey. It is the paper that state officials regard as must reading." Packed with state, regional, and local news, the *Star-Ledger* lives up to its reputation as one of the biggest and best newspapers in New Jersey.

Nine of New Jersey's 26 newspapers serve the northern portion of the state. These papers help residents and newcomers alike to keep abreast of local and national news, traffic, weather reports, and sports in their respective areas.

Morris County's *Daily Record* leads the group. The *Daily Record* maintains an enviable reputation in the heart of Morris County. Its excellent local and state reporting distinguishes it from many of its competitors. In November 1987, ownership passed from the Tomlinson family, which owned the paper since its founding 87 years ago, to Ingersoll Publications Company of Princeton. Other dailies in the north include the *Daily Advance,* the *Daily Journal, The Dispatch,* the *Herald News,* the *Jersey Journal,* the *New Jersey Herald,* and *The News of Paterson*. Each in its own way has maintained the special flavor and uniqueness of its communitites and readership. Many fine non-daily newspapers also serve each locality and their particular needs. In Bergen County alone, there are 45 such non-dailies. The *Suburban News* and the various "shoppers" represent only a few that have gained immense popularity. Featuring news about food, retailing, and community, these publications have acquired a strong and loyal readership.

The offset presses at **The Record** *were custom designed to allow four-color photos to be printed on 16 pages, and 48 pages of spot color. Courtesy, The Record*

These computers play a significant role in every aspect of **The Record's** *production. Courtesy, The Record*

TELEVISON AND RADIO New Jersey residents are twice blessed: they can avail themselves of every New York television and radio station, and they can enjoy their own fine stations as well. In northern New Jersey, they include WCBS-TV (channel 2) in Jersey City; WNET-TV (channel 13) in Newark; WNJU-TV (channel 47) in Newark; WWHT (channel 68) in Newark; WXT (channel 4) in Secaucus; and WWOR-TV (channel 9) also in Secaucus. With the addition of channel 9, New Jersey established its only VHF television station.

Originally licensed in New York, WOR became a New Jersey station as a result of legislation sponsored by Senator Bill Bradley (D-N.J.). The license was transferred to New Jersey in April 1983, although its studios did not commence the physical move from 1440 Broadway in New York to the beautiful new broadcasting facility in Secaucus until 1986. The history of the station, however, begins in New Jersey in 1922 when R.H. Macy, the retail entrepreneur, established a radio station as part of a promotion to dispose of surplus radio equipment in his Newark-based Bamburger's store. Eventually the radio station became a television station with the advent of the new medium. In 1952, RKO bought the station. Since April 1987, WOR-TV has been owned by MCA Broadcasting, a subsidiary of MCA Inc. MCA is one of the largest entertainment conglomerates in the world. It owns MCA Records and Universal Studios, which produce television programs and movies. It also owns the Spencer Gifts store chain. The $387 million purchase price was the second highest ever paid for a station.

Since its address and name change, WWOR has adjusted its programming. The station's community outreach projects have been lauded for their benefit to the public. The New Jersey Associated Press Broadcasters Association has recognized the station's comprehensive news coverage by deeming it the best newscaster of 1986. The station's sports coverage continues to be first-rate, acting as the home station for the Knicks' basketball games and the Rangers' hockey games, besides many sports specials. WWOR-TV has also been the station for the New York Mets baseball games for 26 years.

Sixty-two commercial radio stations broadcast in New Jersey as well. Each has a particular format designed to titilate local listeners. There are six stations in Bergen County alone, with 17 more spread throughout Passaic, Sussex, Morris, Hudson, and Essex counties. WPAT-FM in Clifton is one of the area's leading stations, and it successfully competes with the colossal stations in the metropolitan market.

WWOR-TV9 in Secaucus was recently acquired by the MCA Broadcasting Company for $387 million. Photos by Rich Zila

MAGAZINES Northern New Jersey is in no way bereft of quality magazines. Many top-notch publications circulate throughout the region, and help to keep people aware of trends, events, entertainment, food, personalities, business news, etc. No topic is left unexamined in this market, which is marked by large numbers of affluent, sophisticated, and selective readers. From general purpose to the narrowly focused, northern New Jersey's storehouse of magazines contain every color in the journalistic spectrum.

Commerce Magazine, the official publication of the Commerce and Industry Association of New Jersey in Hackensack, serves as an example of a publication geared to meeting the informational needs of the top executive.

When it began, the 16-page magazine was published as the *Bergen Magazine.* This year, the slick four-color magazine celebrated its twentieth anniversary. *Commerce Magazine* has expanded to fill 100 pages, and retains its focus on business and related concerns. According to Jim Cowen, the magazine's editor and publisher,

Our magazine is a rifle-shot to topline executives in northern New Jersey. Their concerns are ours, and we do our best to provide comprehensive articles on legislative trends, the economy, foreign trade, new construction, and new services. We are also mindful of our duty to help business people to keep abreast of legislative and regulatory proposals for industry.

The magazine also includes historical pieces on the region, and articles on people who have been instrumental in its growth.

Among the numerous publications circulated in northern New Jersey there are several others of particular note because of their high quality. *New Jersey Business,* now in its thirty-fourth year, has withstood the test of time to become a fixture for most astute New Jerseyans. The magazine is published monthly by New Jersey Business and Industry Association in Fairfield. Donald L. Hahnes is vice president/publisher and James T. Prior is assistant vice president/editor. Regular features of interest to businesspeople include news on business and industry, real estate, banking, and advertising; corporate close-ups and in-depth interviews complete its coverage.

New Jersey Monthly in Morristown, with over 340,000 readers, circulates to all the state's counties. This glossy publication covers it all, from the arts to investigative and political writing. The magazine holds something for anyone interested in New Jersey. It is published by Aylesworth Corporation.

New Jersey Goodlife, published by Lifestyle Media Group, Inc., in Somerset, distributes to select households in nine northern New Jersey counties. Its polished and patrician appearance is unique and lives up to the allusions its name conjures. Its 10 yearly issues are privately circulated and carry articles of local and national interest about fashion, travel, food and drink, homes and interiors, and entertainment. The end result is a magazine unlike any in the state.

In a market largely dominated by the New York media, northern New Jersey's newspapers, television and radio stations, and magazines have not merely played "second sister" ; they have created products that are specifically sculpted to suit the highly-educated, sophisticated, and affluent suburbanite.

Northern New Jersey major utility companies provide local residents with sophisticated technology and quality service. Photo by Tom King

With little fanfare, northern New Jersey's major utilities quietly and efficiently do what they do best: provide the region with the resources and communications vital to its existence. Within a complex infrastructure of pipes and wires, energy and information provide support for the area's spectacular success. Both business and residential consumers alike share in the benefits of one of the nation's most sophisticated information and utility networks.

HACKENSACK WATER COMPANY   Hackensack Water Company has gained recognition as one of the nation's largest water purveyors. Its history is intertwined with the region it has served for over 100 years. From its incorporation in 1869 to the present, Hackensack

CHAPTER
SEVEN

Water Company (HWC) has kept in stride with the rapid growth of Bergen and Hudson counties. Providing water to more than 70 towns and communities in the northeast, the firm distributes 100 million gallons of water each day to more than 800,000 residents.

As the largest investor-owned water utility in New Jersey, HWC began its unique history of service just after the end of the Civil War.

Providing The Basic Services

The Hackensack Water Company supplies water to a population of approximately 800,000 individuals living in 60 municipalities within north Hudson and Bergen counties. Photo by Rich Zila

Even then, the New York commuter found Bergen County an agreeable place to live. With business flourishing in the post-Civil War boom, natural resource needs became a paramount concern. Without a public water supply, Bergen's 4,000 residents relied on shallow wells for water. Two men sought to rectify the situation by obtaining water company charters. Garret Ackerson, Jr., a Democrat, and Charles H. Voorhis, a Democrat turned Republican, extended their political rivalry by vying for the same goal. In the end, Voorhis managed to buy out Ackerson only to see the company crash into bankruptcy after the Panic of 1873. However, new investors rescued the neophyte company from financial disaster and nurtured its growth until it became one of the country's largest water utilities.

Today, 14 billion gallons of water are collected in four reservoirs from the crystal-clear headwaters of the Hackensack River to supply customers in New York as well as New Jersey. The company also maintains two large purification plants in Haworth and Oradell. United Water Resources (UWR), a holding company created in 1983, is the parent company of Hackensack Water Company and its subsidiary, Spring Valley Water Company, which services most of Rockland County, New York. UWR, which was formed through a restructuring of the water company, also operates several other water-related enterprises, including Rivervale Realty Company and Laboratory Resources.

WANAQUE SOUTH PROJECT   According to Robert A. Wiener, public affairs representative for Hackensack Water, the company has "grown to serve the needs of the community." That growth is visible in several significant projects that make a priority of meeting the tremendous future demand for water. HWC has ensured the system against the threat of drought well into the twenty-first century, according to Wiener, with the creation of the Wanaque South Project, a complex of pipelines and pumping stations and a major new reservoir at Monksville.

The severe drought that plagued the eastern portion of the United States in 1980-1981 gave impetus to the project, which represents a partnership between the public and private sectors. (HWC and North Jersey District Water Supply Commission are co-owners.) Heralded by public officials and private citizens alike, Wanaque South will boost regional supplies by more than 80 million gallons a day, even under

drought conditions. In its third-quarter report, Robert A. Gerber, president and chief executive officer of United Water Resources, called the dedication of the Wanaque South Project an "important milestone, completing the drought-proofing of Hackensack Water Company."

OZONE TREATMENT   Although quantity is important, HWC is equally committed to delivering a product of the highest quality. To that end, the water company conducted a five-year study to test the acceptability of converting to an ozone system of purification. As a result, HWC is converting and expanding its Haworth Filtration Plant to employ the new ozone purification process. Ozone, a naturally occurring gas discovered in 1840, has been used in Europe for water purification for 60 years. In our atmosphere, ozone screens out the destructive rays of the sun; when added to water, this colorless, orderless gas acts as a potent disinfectant. The process ensures quality water free of the unwelcome by-products of chlorination. With the installation of the ozone process, HWC becomes one of the first water companies in the United States to utilize this new technology.

Newly placed "lifts" of Roller Compacted Concrete must be kept moist and clean, despite the use of heavy equipment. This section is being sprayed. Courtesy, Hackensack Water Co.

The Hackensack Water Co. is investing in the expansion of the Hayworth Water Treatment Plant. This investment will command the use of ozone as opposed to chlorine as a disinfectant. These glass tubes are used to facilitate the treatment process. Courtesy, Hackensack Water Co.

HOMER   Hackensack Water Company's interest in finding ways to better please its customers is reflected in its latest innovation: the Hands-Off Meter Reading system, better known by its acronym HOMER. The first water company in the country to use such a system, HWC will complete the conversion of its entire system within three years. HOMER is the result of more than a decade of research, planning, and development.

With this new technology, HWC can obtain an automatic meter reading through telephone lines without having to go to individual homes. By switching to this automated system, HWC eliminates the inconveniences and problems of the "estimated bill," and the automatic reading will not interfere with normal phone service. A computer dials each phone using a special New Jersey telephone circuit that will not cause the phone to ring.

RIVERVALE REALTY   In 1983, Rivervale Realty became a reality when 700 acres of non-utility land was transferred from Hackensack Water Company to the realty company, a landholding subsidiary of

United Water Resources. The transfer was the culmination of years of study prompted by the New Jersey Board of Public Utilities, which urged the water company to take an inventory of its property. With the advent of new technologies in environmental controls, it was determined that 700 acres of water company land was outside the land-buffer requirements necessary to protect water quality in reservoirs and wells.

Today, Rivervale Realty owns approximately 1,000 acres of undeveloped land in Bergen and Rockland counties. Its goal is to see those lands developed in a phased, environmentally sensitive manner, and to buy, manage, and sell other properties. According to Lillian Ciufo, real estate manager for the realty firm,

We emphasize the environmental aspect, and that makes us unique. We are constantly in contact with local officials to make sure we meet state regulations and that the properties are totally investigated. In fact, we have a full environmental management team that looks at every aspect: wetlands, air, noise, traffic, soil, site history, water quality, and test results. We work at being thorough because it's important to us.

The Monksville Dam "spillway" is 200 ft. wide and has 59 steps. These steps eliminate the need for large energy-dissipating structures normally required with a conventional system. Courtesy, Hackensack Water Co.

UNITED WATER RESOURCES

The 150-foot-high, 2,500-foot-long Monksville Dam holds back 7 billion gallons of water in the new Monksville Reservoir.

Few areas of the United States have experienced such dramatic growth in population and economic enterprises as the northern New Jersey region since the end of World War II. Many conditions were present to foster this growth—entrepreneurs, financial resources, available land, constructive tax structure, and the encouragement of governmental agencies—to name a few.

In addition to those elements, the region has been blessed with a most elemental resource in determining the efficacy of an area to sustain life and, ultimately, economic activity. That resource is the fundamental prerequisite for life: water.

From the days of the native Leni-Lenape to those of the first Dutch and English settlers, the waterways and ample rainfall in what is now northern New Jersey were extremely attractive for farming as an economic activity, as well as for the quality of life in the region. The growth of Bergen County, especially during the past 100-plus years, has been determined by the ready availabil-

ity of the liquid of life—in large part through the efforts of the Hackensack Water Company.

In many ways the growth of Bergen County and that of the water company are symbiotic, for the expansion in population, industry, and commerce

An architect's rendering of the Haworth Ozone plant currently under construction. The completion date is set for late 1988. The plant will have a capacity of 200 million gallons of water a day and will become the sole treatment plant for Hackensack Water.

could not be sustained without the modern public water utility. Now, as the twentieth century nears its end, new challenges and opportunities arise that require imaginative ways to handle the public's need for water while also creating a stronger financial entity for shareholders.

Constructed on the strong base of the Hackensack Water Company, United Water Resources of Harrington Park was formed in 1983 as a holding company. While today committed to providing ample water to Bergen, Hudson, and Rockland counties through two water utilities—Hackensack and Spring Valley water companies—United Water also is actively pursuing other, nonutility businesses.

Today United Water Resources is a profitable, well-balanced, diversified, publicly held corporation organized into two groups: public water utilities, and real estate and other diversified activities.

One subsidiary, Rivervale Realty Company, is engaged in real estate acquisitions, development, and sales. The company's inventory includes 1,000 acres of undeveloped land in the midst of perhaps the richest real estate market in the nation—averaging only 15 miles from midtown Manhattan.

Closely aligned with United Water Resources' original business—namely supplying water—are Laboratory Resources, Mid-Atlantic Utilities, Metering Services, Inc., and a broad-based team exploring possible business

Rivervale Realty Company, a subsidiary of United Water Resources, will build this seven-story, 200-condominium development in Fairview, New Jersey, on the site of a former open reservoir.

opportunities in water-related fields.

Laboratory Resources is a commercial laboratory providing the most modern environmental services to water and wastewater utilities, as well as a host of other businesses ranging from oil and petroleum companies to chemical manufacturers and power and light utilities.

The firm's broad expertise includes drinking water and wastewater technology, treatment, and management; chemical analysis; environmental consulting services; and laboratory sales, marketing, and customer services, which enable its customers to comply with broad environmental compliance regulations or provide improved water supplies and services.

Mid-Atlantic Utilities provides water and sewer system design, construction, and operation for new real estate developments based on the corporation's more than 100 years of experience in these fields, dating from the founding of the Hackensack Water Company in 1869.

Metering Services, Inc., markets the

firm's expertise in automatic meter-reading systems to other utilities.

United Water's Business Development Center provides research into other water-related businesses that the company might enter, including services to water utilities outside the area.

It remains the supplying of water through its Hackensack Water Company unit, however, that is United Water's principal service to the northern New Jersey market. The utility has expanded its reservoir capacity by its participation in the Wanaque South Project and, thereby, its ability to become "drought-proof." The water company's objective is to safeguard its service area from the economic, health, and comfort costs of restricted water availability and use, due to both natural causes and man-made environmental hazards.

The demand for water is growing.

During 1987 the peak day demand reached 148 million gallons—nearing the record 160 million gallons for one day set in 1980 during a severe heat wave. The Wanaque South Project, the expansion of the Haworth filtration plant and its conversion to the ozone treatment process, plus improvements in pumping, filtration, and distribution systems are major efforts toward meeting the need for increased water supplies of a high quality.

United Water companies also seek new ways to improve service to customers and to reduce costs. One of these innovations is the computerized HOMER (Hands-Off Meter Reading) system, which enables the reading of water meters over telephone lines without affecting telephone service, and that eliminates estimated bills forever.

Although United Water Resources has broadened its base to include real estate development and other nonutility activities, water is still the company's middle name and the principal thrust of its endeavors as it moves toward the twenty-first century.

NEW JERSEY BELL

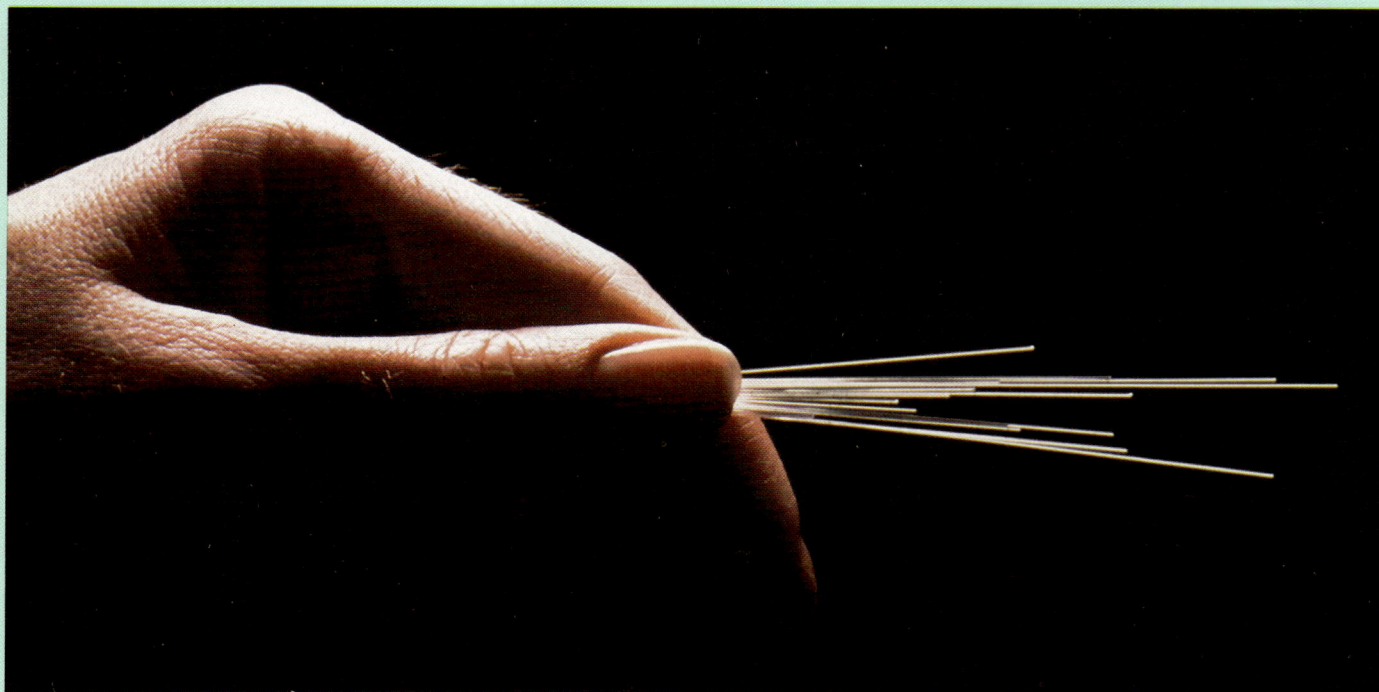

Fiber-optic cable being installed by New Jersey Bell is paving the way for the introduction of innovative new services. At the close of 1987, 63,000 miles of fiber-optic cable had been installed statewide.

New Jersey has proven it has the right stuff, and New Jersey Bell aims to keep it that way.

Since 1927, when the Delaware and Atlantic Telegraph and Telephone Company changed its name to New Jersey Bell, the company has worked to make New Jersey strong. During that first year, when 600,000 telephones were connected to the firm's network, New Jersey Bell adopted principles of quality and service that it has adhered to ever since. As part of Bell Atlantic, New Jersey Bell's approximately 20,000 employees continue to build on that commitment to quality.

New Jersey Bell has worked to make the Garden State a more attractive place in which to work and live. And the company is proud that its telecommunications services are part of the reason people and businesses choose to locate in New Jersey.

Providing high-quality service at a low price comes from skillful deployment of technology, product innovation, a commitment to service, and sound management. As part of Bell Atlantic, New Jersey Bell uses the expertise and product offerings of its sister Bell Atlantic companies, responding fully to customer needs with an expanded package of systems and ser-

vices. This includes telecommunications services and equipment, computer system maintenance, and financing.

Through joint marketing with Bell Atlanticom Systems, Inc., which sells telecommunications equipment, both large and small business customers can enjoy the convenience of one-stop shopping with New Jersey Bell—the ability to buy telecommunications services and equipment from one source. This allows New Jersey Bell the chance to offer commercial customers a more integrated approach to meeting their communications needs.

With the lowest basic residential and short-distance toll rates offered by any Bell operating company in the nation, New Jersey Bell helps customers keep expenses low. That, in turn, is good for the economic health of the firm's 2.8 million residential and 432,000 business customers, who are enjoying a five-year cap on basic local service and toll rates. That means customers will be getting service in 1993 at 1985 prices.

New Jersey Bell is able to keep costs

low by actively managing technology and putting it to work in its statewide network. More than $600 million went into capital improvements in 1987, with plans to invest a similar amount in 1988.

At the close of 1987, 63,000 miles of fiber-optic cable had been installed statewide, and 179 of the company's 213 switches were electronic or digital electronic—paving the way for the introduction of innovative new services.

For example, New Jersey Bell is conducting a trial of seven advanced calling features called CLASS Calling Services in Hudson County and Atlantic City. One CLASS feature, Caller ID, lets you know the number of the caller before you answer the telephone. Another feature, Return Call, automatically dials the number of the last incoming caller, even if you did not answer the call. The company hopes to introduce CLASS statewide beginning in 1989.

Other trials in Holmdel and Red Bank are testing ISDN, the Integrated Services Digital Network. ISDN is an all-digital network that will allow customers to send and receive voice, data, and image signals simultaneously over existing telephone lines. When fully implemented in the 1990s, ISDN technologies will make it possible for cus-

With 179 of the firm's 213 switches electronic or digital electronic, New Jersey Bell is preparing for the introduction of innovative services such as Integrated Services Digital Network (ISDN), an all-digital network that will allow customers to send and receive voice, data, and image signals simultaneously over existing telephone lines.

tomers to access those various signals as easily as they make a telephone call today.

St. Joseph's Hospital in Paterson is reaping the benefits of a Central Office-based Local Area Network (CO-LAN). Through a new application of existing network technology, New Jersey Bell is using existing wiring to provide simultaneous voice and data transmission among a number of hospital locations.

Personal computer users statewide can access listings of New Jersey restaurants, complete with information about menus, prices, and ambience, thanks to some creativity and New Jersey Bell's Public Data Network. The Public Data Network allows customers to send and receive information easily and inexpensively through the use of a common data network. The restaurant information service, called Dine Out, is offered to personal computer users throughout northern New Jersey for the cost of a local call.

Internally the firm also introduced several software systems to provide faster and more efficient customer service. For example, one system eliminates the manual writing of service orders when customers order or change service. In addition to speeding the order entry process, the system quickly routes information to the installation and other departments that complete the order.

With some regulatory restraints recently lifted, New Jersey Bell will test new information services, including message answering, voice mail, and a coin phone message system in late 1988.

Aimed at residential customers, message answering service works like an answering machine, but requires no equipment in the home. The subscriber records a personal greeting, which is played for callers when the subscriber's line is busy or is not answered within a certain number of rings. Callers then may leave a message, and a special dial tone or other signal will notify the customer that a message is waiting.

A similar but more advanced service, called voice mail, is expected for businesses. This will allow them to have their phones answered 24 hours a day without the need for any equipment except a standard telephone. With this service, callers also will be able to forward their calls and assign priorities to their messages.

Another service to be tested in 1988 will allow a coin phone customer to leave a message for someone who cannot be reached. The message will be delivered later when the person is available.

Through this constant program of developing innovative services and delivery systems, New Jersey Bell continues to provide the quality service and products that meet business and residential customers' ever-growing telecommunications needs.

New Jersey Bell's commitment to excellence in both quality and service has remained steadfast for the past 60 years. As part of Bell Atlantic, New Jersey Bell's nearly 20,000 employees continue to build on that commitment to quality.

AT&T Bell Laboratories is one of the world's leading research centers. Since 1925 it has averaged one patent per day. Here, we see the interior of an anechoic (echo-free) chamber. Photo by Bob Krist

This Coast Guard vessel was on hand in Elizabeth to combat a chemical dump fire and assist with cleanup activities. Photo by Bob Krist

Ciufo underscores the fact that Rivervale's environmental priorities are concomitant with HWC's philosophy: "Our interest in preserving the integrity of the land coincides with the water company's longstanding mission to fulfill its responsibility to society and the physical environment." Rivervale is doing just that in its plans for Lake Forest Corporate Park, a combined office-residential complex that will occupy a 160-acre wooded site in Emerson. "Seventy percent of the land will be left in its natural state," Ciufo explains. "For example, there are beautiful cottonwood trees on the property which go up 100 feet. These trees are hundreds of years old, and we would never dream of destroying them. We work around natural features to preserve the natural landscape."

Lake Forest Corporate Park will have 900,000 square feet of office space and 200 condominiums when it is completed. In explaining Rivervale's success, Ciufo says, "Each municipality and town knows we will work with them to enhance what exists. After all, we have been corporate 'neighbors' through the water company for over a century." Ciufo adds, "I live in Bergen County, so I fully realize that every town is beautiful and unique. Each has some special feature or aspect, and I like feeling that we preserve and contribute to that."

PUBLIC SERVICE ELECTRIC & GAS With a service territory that extends diagonally across the state from the Hudson River in Bergen County to the Delaware River in Gloucester County, Public Service Electric & Gas (PSE&G) provides natural gas and electricity to seven out of ten residents of New Jersey. It is the largest electric and gas company in the state, and the third-largest combined utility in the nation. Covering 2,600 square miles, PSE&G is the premier utility in the state's industrial and commercial corridor, which includes six major cities and nearly 300 suburban and rural communities. More than 2 million customers depend on the firm for electric and gas service.

The firm originated as an extension of the first central generating station devised by Thomas A. Edison in 1882. The organization that was to become PSE&G was founded in 1903 by a group led by Thomas N. McCarter. McCarter's group brought together numerous railway, gas, and electric companies, and since those early days, PSE&G has

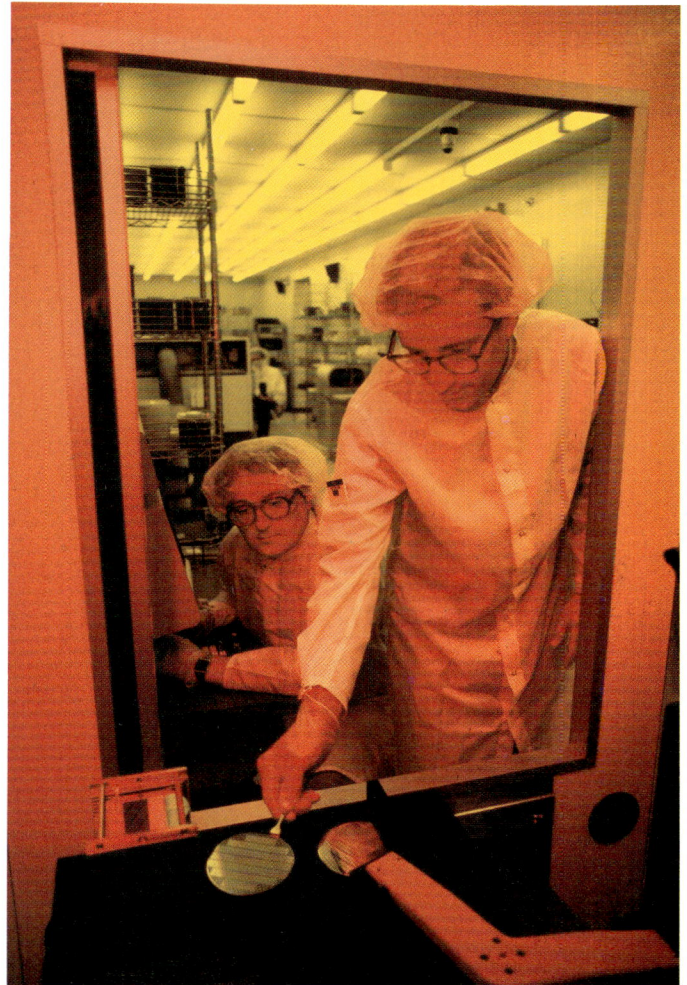

Recent developments at the AT&T Bell Laboratories include the one-megabit computer memory chip. Photo by Bob Krist

grown tremendously. Today, the company possesses an electric generating capacity of 10 million kilowatts. Its average daily gas capacity is approximately 20 million therms. PSE&G's most famous and visible recipient of electricity is a national treasure: the Statue of Liberty. The company has kept her well lighted since 1916.

PSE&G is a member of a power pool known as the Pennsylvania-New Jersey-Maryland Interconnection, which consists of 11 operating electric utilities. The use of reserve pooling is cost effective for the customer because it relieves each individual company from maintaining larger reserve capacities of its own. The transmission network also provides hourly economic transfers of energy among member companies, and economic import of energy from neighboring regions. In that way, PSE&G and the other members can take advantage of excess coal energy available in eastern regions to reduce oil consumption. Each year, the PJM power pool saves PSE&G customers millions of dollars.

Throughout its history, PSE&G has been a prime contributor to the state's phenomenal economic growth. The company's vital and aggressive Area Development Department provides a treasure chest of information for interested inquirers. By producing a plethora of free in-

formational materials, providing client consultation, and conducting site tours, PSE&G's market specialists have assisted many firms to locate in New Jersey. Site-finders conduct personalized tours and provide detailed descriptions of the properties in terms of labor supply and rates, taxes, availability of utilities and services, transportation, educational and recreational facilities, business climate, building costs, and housing availability.

NEW JERSEY BELL New Jersey's substantial contribution to communications dates back to Thomas A. Edison's developments and improvements in telephone transmission. Conducted for his friend, Alexander Graham Bell, Edison's successful experiments made the telephone a feasible item for use by the general public.

New Jersey Bell has kept that tradition of experimentation alive. As the firm that pioneered telecommunications in the state, its sales and engineering teams continue to develop products tailored to the needs of both businesses and the general public. In an increasingly service-oriented economy, New Jersey Bell constantly seeks new means by which to help businesses exchange information easily. In 1986, the firm introduced more new services than in any other year in its long

ROCKLAND ELECTRIC COMPANY

Serving a dynamic region's electric needs, Rockland Electric Company, the New Jersey subsidiary of Orange and Rockland Utilities, Inc., plays both an important part in the daily lives of its customers and a significant role in the economic vitality of its service territory in Bergen, Passaic, and Sussex counties.

The northern New Jersey region that Rockland Electric serves enjoys a robust economy, with continued growth in both the residential and commercial sectors. The firm has taken steps not only to meet the region's growing energy needs, but also to encourage its continued economic expansion through attractive rates and an aggressive area development effort.

"As part of this community," says chairman and chief executive officer James F. Smith, "Rockland Electric plays a vital role in encouraging the economic growth of this region. As a result of this effort, our customers enjoy the direct benefits of an enhanced business community and the savings generated through a broadened rate base."

To accommodate the projected growth of its service territory, Rockland Electric has developed a long-term energy plan to make optimum use of its

The region served by Rockland Electric Company continues to enjoy a healthy economy, and the firm's commercial department works closely with area developers to ensure that the energy needs of the community are met.

Rockland Electric Company's parent company, Orange and Rockland Utilities, Inc., has completed the $120-million reconversion of its Lovett Generating Plant on time and on budget. The plant can now choose among coal, oil, or natural gas as fuels to generate electricity.

existing electric generating facilities, along with other available energy sources, as a means toward maintaining service at an affordable cost while avoiding the need to construct new power plants.

According to Thomas A. Griffin, Jr., president and chief operating officer, an example of this effort is the company's recent $120-million coal reconversion project at its Lovett Generating Station on the Hudson River in Tomkins Cove, New York. "This project gives us the flexibility to use whatever fuel is most economical and available to generate electricity (oil, gas, coal), resulting in fuel cost savings for our customers today and tomorrow," he explains.

Orange and Rockland's Master Energy Plan also drives an aggressive conservation effort, from conducting industrial energy management seminars, marketing efficient heating and cooling equipment, sponsoring home weatherization workshops for civic groups, to conducting rate incentive experiments. Its strategy is to accommodate increased overall energy consumption while leveling demand "peaks" that could signal the need for new generating

facilities.

Rockland Electric Company has served northern New Jersey since 1898. "As a longtime member of the community," says Linda Winikow, vice-president/corporate policy and external affairs, "we are dedicated to continuing our tradition of quality service while always looking for new ways to be a 'good neighbor' to our customers."

According to Mr. Smith, Rockland Electric's success in responding to its community's changing needs has benefited the company's shareholders as well as its customers. Through prudent planning, the firm has achieved stability in its electric rates while also establishing a very sound capital structure attractive to investors. He notes that the company has continued to diversify modestly in the nonutility field as another means toward further enhancing its shareholders' investment.

Orange and Rockland Utilities, Inc., and its subsidiaries serve an area of 1,350 square miles and a population of more than 640,000 in northern New Jersey, southeastern New York State, and northeastern Pennsylvania.

The Network Operations Center of AT&T, which is the computerized control facility for the U.S. long-distance network, is based in New Jersey. Photo by Bob Krist

history, including Public Data Network (PDN), Central Office-based Local Area Network (COLAN), and Centrex III Community Centrex.

PDN allows any type of computer or terminal to communicate with virtually error-free reliability. There is no need to lease private lines or buy costly conversion devices. As a result, there is easy access to data bases, electronic home banking and shopping, videotext, and computer bulletin boards. The service is also ideal for companies that need to transmit data such as payroll information, or for businesses that need to secure charge card or check approvals.

COLAN allows simultaneous voice and data transmission over existing telephone lines, eliminating the need for costly dual facilities. It provides full digital connectivity between independent data devices such as mainframe computers, terminals, PCs, and telephones.

Centrex III provides the convenience and feaures of a Private Branch Exchange (PBX) without the need to make large capital expenditures for equipment that can become obsolete. The service was developed for use by the shared tenant service providers.

According to its 1986 annual report, New Jersey Bell offers its residence customers the lowest rates of all such customers served by Bell operating companies in the United States. Its business rates and intrastate toll rates are also among the lowest in the country.

During 1986, New Jersey Bell invested more than $580 million to improve and maintain its statewide telecommunications network.

ORANGE AND ROCKLAND UTILITIES, INC. Orange and Rockland Utilities and its subsidiaries—Rockland Electric Company

and Pike County Light and Power Company—provide electric service within a 1,350 square-mile territory. The areas served include Bergen, Passaic, and Sussex counties in northern New Jersey; Orange, Rockland, and Sullivan counties in New York; and Pike County in Pennsylvania. The firm also provides natural gas in a substantial portion of its service territory.

The utility foresees stability in rates through the end of the decade due to several significant factors. One of the most important of these is the advantage Orange and Rockland has in being the only utility in the northeast with no nuclear power investment. It therefore avoids the potential for excessive cost overruns. The reconversion of the Lovett generating plant from oil to low-sulfur coal also insulates customers from rate increases.

Orange and Rockland's area development efforts extend to New Jersey as well, with over one million square feet of available office space currently in inventory in the New Jersey communities of Mahwah, Montvale, Ramsey, and Upper Saddle River.

Northern New Jersey utilities provide the natural resources, power, and communications capabilities necessary to support the region's explosive growth as a corporate haven. Their advantages offer an attractive lure for companies contemplating relocation. Rich in water—long considered one of the state's most important resources—northern New Jersey's reservoirs more than meet the considerable needs of the region. Its bountiful supply has been assured well into the twenty-first century. Gas and electric companies amply and dependably meet the energy needs required to fuel daily life. By keeping rates down, these utilities have attracted New York City firms seeking the savings to be found by doing business in northern New Jersey. In communications, the region is similarly unsurpassed. By offering the innovative communications services paramount to an information-service economy, northern New Jersey assures its economic leadership now and into the future.

Local ballet companies consistently raise the standard of New Jersey's contribution to the performing arts. Photo by Aram Gesar

Who knows the Palisades as I do knows the river breaks east from them above the city—but they continue south —under the sky—to bear a crest of little peering houses that brighten with dawn behind the moody water-loving giants of Manhattan.

Photo by Michael Spozarsky   William Carlos Williams

These lines appear in a compilation of selected poems by one of America's most seminal poets. A physician practicing in Rutherford, William Carlos Williams was born and remained in the northern New Jersey town throughout his life, deriving his poetic inspiration from its people and environs. As Paul Mariani says about Williams. "His was a life lived by choice in the provinces attending the needs of the people he came up against in Rutherford, Hackensack, Passaic, Carlstadt, Lyndhurst, Garfield, Paterson." Williams knew instinctively that "place" was the only reality, and as he asserts in the *Quarterly Review of Literature,* "we do not have to abandon our familiar and known to achieve distinction." The high praise Williams' oeuvre receives

CHAPTER
EIGHT

from literary critics 25 years after his death proves that he was right.

But just as it took northern New Jersey's "homegrown" poet years to convince the world that a major talent could sprout from American

A New Venue For The Arts

literary soil, so, too, has it taken the area itself years to attain a reputation as a center for the arts. Traditionally, it has provided audiences for cultural events in Manhattan, its neighbor across the Hudson River. However, the huge success of the Meadowlands Sports Complex, the enormous business and industrial growth in the region and state, and an aversion, on the part of northern New Jerseyans, to battling city traffic, has contributed to new construction, renovation, and the expansion of arts facilities throughout the area. This has helped northern New Jersey come into its own as a thriving cultural center. In turn, its phenomenal cultural growth in the arts has contributed to enriching the economy. According to a study released by the Port Authority of New York-New Jersey, an investment in the arts is a very real one because for every dollar spent on the arts, four are fed into the economy. An acceleration of its cultural activity has also given northern New Jersey a quality of life that is surpassed by few other places in the country, making it an attractive choice for corporations contemplating a move to the region.

The William Carlos Williams Center in Rutherford was named for the hometown, Pulitzer Prize-winning poet and physician. Photo by Sharon Sullivan

Oktoberfest is one of the many celebrated festivals held at Waterloo Village in Stanhope. Photo by David Greenfield

CULTURAL CENTERS AND THEATERS When William Carlos Williams confessed "Here [in Rutherford] the world of art is nonexistent," he had no way of knowing what the future would hold. In his beloved town, the William Carlos Williams Center for the Arts now stands near his home, which has been declared a historic monument. Located only minutes from the Meadowlands Sports Complex, the center is easily accessible by various routes.

The center, northern New Jersey's largest multi-disciplinary arts facility, is built around—and under—the Rivoli, a vaudeville theater constructed in 1922 and damaged by fire in 1977. The Rivoli will eventually be restored as the center's 1,550-seat concert hall. The theater, which is a combination of classic-rival and baroque style, was designed by Passaic architect Abram Preiskel. Its value was recognized when it was designated a Historic American Theatre. Its distinctive decorative elements stand out as symbols of the painstaking craftsmanship of a cherished era. Fortunately, the Rivoli's spectacular original chandelier survived the 1977 fire intact. Weighing two tons, it is 8 feet wide and 12 feet high. A twin of the one in the old Warner's Theater in New York, the chandelier contains 62,000 prisms of hand-cut and hand-polished Czechoslovakian crystal. The restored fixture, valued at $125,000, now hangs awaiting the completion of the theater's renovation.

According to former executive director Victoria Hardy, the theater is not merely a valued remnant of the past, it is a practical showcase for future performances: "The stage is 31 feet deep and 64 feet wide and has a proscenium arch 36 feet wide and a grid six stories

high. This will enable us to provide performances of opera, theater, orchestras, and dance companies."

The William Carlos Williams Center for the Arts is committed to showcasing for a local audience native talent in several artistic disciplines, including poetry, music, dance, and theater. The center is also host to New York artists. On its current roster, the center includes a broad spectrum of artistic offerings to tempt the cultural palate. In the tradition of its namesake, the center presents poetry readings, workshops, and seminars. It has also taken the lead in presenting a varied menu of musical performances—from Gershwin to Beethoven, from New Jersey composers to international ensembles. Over 20 music events are presented each month.

Five theater companies utilize the arts center; it produces other independent shows as well. Dramatic readings, stage productions, and backers' auditions are offered. The center's versatile programming also includes a highly innovative and successful children's theater; a classic film series; art shows; outdoor concerts; crafts exhibits and dance performances on the plaza; a paved and planted area adjacent to the center; and a gallery with rotating exhibits highlighting New Jersey artists. In addition, the center houses a wide variety of adaptable areas for rent by businesses and organizations. The Poet's Cafe Sushihara completes the aura of uniqueness captured by the center. Classical Japanese cuisine served in an elegant teahouse setting is available for lunch, dinner, and pre-theater specials.

Bergen County is fortunate to have within its borders another major forum for the arts. The John Harms Center For The Arts in Englewood is also the result of the creative imagination of an art aficionado. John Harms (1906-1981), known as "the Sol Hurok of Bergen County," began his career in 1948 as an impresario. Through his efforts, a succession of artists came to northern New Jersey, includ-

Over 60 years old, Newark Symphony Hall is currently undergoing massive renovations to keep in step with the rejuvenation of Newark. Courtesy, Newark Symphony Hall

ing Joan Sutherland, Van Cliburn, Eugene Ormandy, Arthur Rubinstein, Vladimir Horowitz, and Leonard Bernstein. During the past decade, the center has maintained its reputation as an artistic mecca in Bergen County. Its superb presentations of classical music programming is unrivaled in the region. Several artistic groups are part of the center or make it their permanent home: the illustrious New Jersey Symphony Orchestra has chosen the center as its Bergen County base during tours; the Bergen Philharmonic Orchestra, formed in 1936 as the Teaneck Symphony Orchestra, now performs regularly at the center; the internationally acclaimed Pro Arte Chorale, founded in 1964, recently joined the center's family; the Creative Theater, Inc., draws children from all over New Jersey, Rockland County, and Manhattan, presenting two to four shows a year for children and their families; the Dance At John Harms program, under the direction of

Nancy Goldstein, provides classes for all age groups, and hosts a Teen Touring Company. In addition to the music, dance, and theater that the center presents, it has an art gallery featuring the paintings, sculptures, collages, and photography of artists from the New York/New Jersey metropolitan area; the exhibit changes monthly.

The John Harms Center For The Arts possesses a unique history. Built in 1926, it was originally known as the Plaza Theater. It served as a movie/vaudeville emporium until it was acquired by John Harms and the board of directors in 1976. By 1986 it received the first annual Englewood Historical Society Award "for making a significant contribution to historic preservation/restoration to an Englewood property of historic value." The center was also given a certificate of commendation by the Bergen County Board of Chosen Freeholders "for exemplary contribution to historic preservation." The interior design evokes

The New Jersey Symphony Orchestra, with conductor Hugh Wolff, performs at Newark Symphony Hall. Photo by Arthur Paxton

the late Victorian Art Deco style of the 1930s. Because of its charm and practicality, the facility is frequently used as a shooting location for commercials and motion pictures such as *Annie Hall* and *The Muppets Take Manhattan*.

The center has been known for its classical repertoire, but it has broadened its scope. Today, it produces events that appeal to all segments of the northern New Jersey population. These include such widely diverse shows as rock concerts and New Jersey Ballet Company performances. Comedians, ethnic festivals, and country-western music programs are also slated for future programming. Despite this, the basic subscription series offered to the public will continue to be solely for classical music and dance programs.

The John Harms Center has also made itself available to many businesses and clubs interested in using its highly versatile space for meetings and receptions.

The Orrie de Nooyer Auditorium in Hackensack has been serving the public for 15 years. Its 60-foot proscenium stage and spacious 1,201-seat house make it ideal for an array of events. Audiences are offered an exciting variety of cultural programming that includes full-length grand operas, dance, musicals, and classical and contemporary music. In 1988, the Civic Music Association of Bergen County will present its annual series of classical concerts at the Orrie de Nooyer Auditorium. The series will include performances by pianist Richard Goode, soprano Dawn Upshaw, violinist Ida Levin, and the Brooklyn Philharmonic with Lukas Foss as music director and conductor and Paul Shaw as pianist.

Newark Symphony Hall, built in 1925 by the Shriners, offers an enormous, yet elegant, structure for performers. Beautifully main-

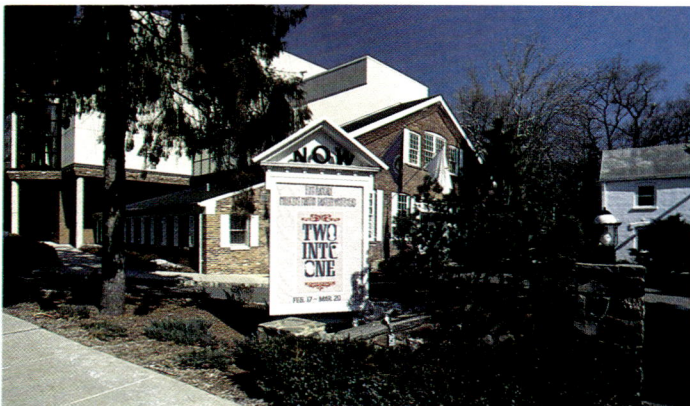

The Paper Mill Playhouse was founded in 1934 by Antoinette Scudder and Frank Carrington. In 1972, the Playhouse was designated the "State Theatre of New Jersey" by Governor Cahill. Photo by Sharon Sullivan

In 1986, the Paper Mill Playhouse presented **Run For Your Wife** *starring David McCallum. The production ran for one month's time. Courtesy, Paper Mill Playhouse*

1940's Radio Hour *appeared at the Paper Mill Playhouse for longer than one month's time. Courtesy, Paper Mill Playhouse*

tained, it rises five floors and contains three underground stories. The stately old theater houses a 2,811-seat auditorium, two 100-seat theaters, a 1,200-square-foot multipurpose ballroom, and two television studios. Considered New Jersey's most opulent and acoustically stunning concert hall, its extensive restoration will make this Old World gem even more attractive. Its renovation is tied into the revitalization of Newark, New Jersey's largest city.

The Paper Mill Playhouse, founded in 1934, is one of America's pioneer regional theaters. One of the nation's largest and most active nonprofit theaters, the renowned Paper Mill in Millburn has a 1,200-seat theater that stages professionally acted Broadway shows, musicals, original theatrical productions, children's programs, dance series, and a summer festival. The Paper Mill is literally enjoying a second life since the fire that leveled it in 1980. Using a $600,000 gift as a starting point, benefactors raised the remaining $5.4 million to complete the theater's reconstruction.

The Paper Mill also has received a prestigious Artistic Focus Grant from the New Jersey State Council on the Arts, which is

designed to assist in developing national and international recognition.

The New Jersey Shakespeare Festival at Drew University in Madison has a long history of excellence. Set amongst the bucolic surroundings of the Morris County university, the festival is unique. An avid theatergoer can enjoy professional level productions during its June through December production season. Classical and modern plays are offered by a professional Actors Equity repertory company. Theater workshops for all ages are also available. During the summer of 1987, the festival celebrated its twenty-fifth anniversary by offering, in nightly rotation, three Shakespearean plays: *The Taming of the Shrew, The Winter's Tale,* and the infrequently performed *The Tragedy of Coriolanus.* The fall menu included the contemporary plays *A Streetcar Named Desire, Present Laughter,* and *Translations.*

Every year, the Waterloo Festival for the Arts in Stanhope attracts

tens of thousands of visitors to the historic village as well as to its musical presentations, which take place on weekends from May through October. Long considered one of Sussex County's treasures, the festival has consistently elicited acclaim for its outstanding and diverse cultural program of classical and popular music, dance, and opera. In 1987, the Metropolitan Opera performed both *Tosca* and *La Boheme* there. Its offerings include a range of other entertainment, from symphonies to world famous soloists. There are crafts and antique exhibitions, jazz programs, and ballet, country-western, folk, and bluegrass performances. Waterloo is a tradition that many are discovering.

The Saddle River Valley Cultural Center in Upper Saddle River was built in 1847. It has been perfectly preserved as a reminder of more pastoral times. "The Little Church on the Hill" today serves the community as a cultural center, offering weekend performances and monthly art shows. Programs include improvisational theater, one act plays, chamber music concerts, piano recitals, swing dance bands, and magic and puppet shows.

New President Donald Schroeder and Chairman Gerald L. Dorf have planned a major expansion of the New Jersey State Opera. This photograph was taken during the performance of Il Trovatore *starring Josella Ligi and Franco Bonanome. Courtesy, New Jersey State Opera*

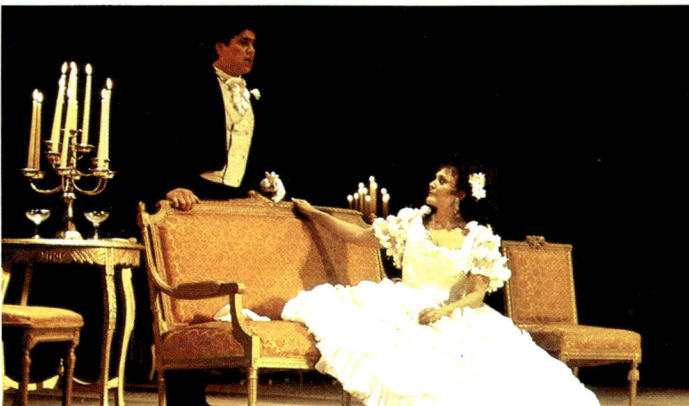

The New Jersey State Opera performs La Traviata *with Adriana Maliponte and Rodolfo Acosta. Courtesy, New Jersey State Opera*

The New Jersey State Opera performs at Newark Symphony Hall. Here, the cast is on stage for one scene in the production of Boris Godunov. *Courtesy, New Jersey State Opera*

MUSIC Long considered New Jersey's prize possession, the New Jersey Symphony Orchestra garners accolades from music critics and arts cognoscenti wherever it performs. The orchestra is the second-largest in the Northeast. It tours throughout the state during its 25-week season, but its permanent home is in Newark. Under the baton of noted conductor Hugh Wolff, it has had one triumphant season after another.

During its tenure as the premier state orchestra, it has upheld a tradition of presenting distinguished guest soloists. The orchestra performs a variety of concerts during the summer on a mobile sound stage, which enables it to reach audiences in difficult locations. It has been said that one no longer has to cross the Delaware or the Hudson in order to hear one of the world's great orchestras—a comment accurate by anyone's standards.

The New Jersey Chamber Music Society, a regional society based in Montclair, draws its members from top area musicians. During its October-to-May performance season, the society plays a repertoire of master works and little-known pieces, both contemporary and classical. In addition, the society has initiated a series of concerts aired on public radio stations from coast to coast.

The Cathedral Symphony of Newark plays works by contemporary composers as well as classics during its season, which runs from September to May. Paul E. Fran is the newly appointed executive director of the Cathedral Concert Series. A musician and an arts administrator, Fran has taught at the Manhattan School of Music. Formerly executive director of the Conductor's Guild, he has also been associate director of Young Audiences, Inc. He has composed scores for television programs and films and is an oboist and English horn player. The symphony gained national acclaim in 1986 when it was the subject of a television special. The 1987-1988 season includes 12 events at the Cathedral of the Sacred Heart in Newark.

Newark's exquisite Symphony Hall is the home of the New Jersey

working exclusively in the classical style. Its intention to advance classical ballet, in the form of nineteenth-century classics and in new ballets that use classical movement vocabulary, is peculiarly unique to the company. That avowed commitment was formulated by Fred and Evelyn Danieli who founded the ballet in 1961, 10 years after creating their ballet school. Through the years, Fred Danieli sought to retain his own link to the dawn of American classical vocabulary through the integrity of the company's performances. Trained under George Balanchine, Danieli was a member of the American Ballet Caravan and Ballet Society, forerunners of the New York City Ballet. He was also a soloist in premieres of Balanchine's "Ballet Imperial" and "The Four Temperaments."

In 1987, Danieli passed the scepter to Peter Anastos, the internationally known choreographer, who assumed the role of artistic director. Anastos has staged ballets throughout the United States and abroad. He has collaborated with Mikhail Baryshnikov, directed opera, and choreographed for television. His credentials and vitality mesh nicely with the company's energetic and artistic projections for the future. As an avid proponent of the classical mode, Anastos and his corps of dancers will launch a production of "Swan Lake" in the autumn of 1988 in collaboration with the New Jersey Symphony Orchestra. The spectacular new production not only reaffirms the Garden State Ballet's serious commitment to classical forms, but it raises the standard of New Jersey's contribution to the performing arts. According to Anastos, "when two of New Jersey's leading cultural institutions join

The New Jersey Ballet was formed in 1958. Since then, the company has worked with the New Jersey Symphony and the New Jersey State Opera on numerous occasions, effectively augmenting its reputation. Photo by Bob Krist

forces on this scale, a much larger statement is made ...and we make it to a much larger public. Additionally, we inform those who look only to New York for great ballet and great music that it thrives here in New Jersey."

The New Jersey Ballet Company in West Orange is best known for its eclectic repertoire of both traditional pieces and original jazz dances. The company travels extensively, bringing the beauty of the dance to locations throughout the state. Since its first performance in 1958, the company has maintained a commitment to offer the highest quality dance to New Jersey audiences and to play a major role in the state's cultural life.

Artistic Director Carolyn Clark is the force behind the company's achievement: it received the 1987 New Jersey State Council on the Arts

State Opera. In its 21-year history, the company has featured some of opera's brightest stars, including Robert Merrill, Placido Domingo, Frederica von Stade, Franco Correlli, and Gilda Cruz-Romo.

DANCE With the recognition accorded northern New Jersey's two ballet troupes, the region's companies emerge forever from the shadow formerly cast by New York's exclusive world of dance. For years, New York dance critics refused to acknowledge the contribution New Jersey companies made to the art. New Jersey's own ballet enthusiasts patronized New York performances while local companies battled for visibility.

However, much has changed. Northern New Jersey's ballet companies have been able to attract formidable choreographers, and this has helped to breathe life into existing works and to create new ones. The state's solid backing of the arts has also helped these companies live up to the potential envisioned by their respective founders. In short, ballet in northern New Jersey is poised on the threshold of an exciting era.

The Garden State Ballet is the only major company in New Jersey

Distinguished Artistic Award for artistic excellence. Clark has utilized her experience with the American Ballet Theater to help catapult the regional company into recognition as a force in New Jersey's ballet world.

The New Jersey Ballet Company has readily collaborated with the New Jersey Symphony and the New Jersey State Opera. Other joint artistic endeavors have swelled the company's already outstanding reputation. Edward Villella has acted as an artistic advisor and choreographer, and Eleanor D'Antuono, a ballerina with the American Ballet Theater, has also been an advisor.

In a more modern approach to dance, the Lillo Way Dance Company of Upper Montclair achieves its artistic goals by utilizing the freedom of experimental works.

ART MUSEUMS AND GALLERIES Both museums and galleries are plentiful in northern New Jersey. You need only to seek, and you shall find every taste represented. The following list represents only a few.

—The Newark Museum possesses a collection that is a lively mix of type and time period. Its collection ranges from decorative arts and American painting and sculpture to Tibetan and Oriental art. The museum supports ancillary branch houses, such as the Ballantine House (a Victorian mansion) and the Newark Fire Museum. The museum also features a planetarium.

—The noteworthy Montclair Art Museum has a collection that focuses on American art of the eighteenth through twentieth centuries. The museum has also amassed an impressive array of fashions and costumes for display.

—The Art Center of northern New Jersey, a not-for-profit art school and cultural center in New Milford, offers a spacious, well-lighted gallery. Prominent New Jersey and metropolitan area artists exhibit at the facility.

—The Old Church Cultural Center in Demarest, originally a church built in 1874, retains its "Carpenter Gothic" character while playing host to monthly exhibits of sculpture, graphics, painting, photography, pottery, and crafts by artists of national prominence.

—The Bergen Museum of Art and Science in Paramus, the only broad-based museum in northern New Jersey, offers exciting and creative temporary exhibitions as well as permanent art and science displays. The museum's fine-art collection focuses on twentieth-century American art.

—The recently reopened Jersey City Museum possesses the city's cache of nineteenth- and twentieth-century paintings and decorative and historical objects, in addition to over 300 drawings, watercolors,

The Newark Museum houses an eclectic selection of artwork. Photo by Carol Kitman

The Montclair Art Museum is noted for an outstanding collection of costumes and apparel. Photo by Sharon Sullivan

The Victorian Ballantine House is located adjacent to and is financially supported by the Newark Museum. Photo by Sharon Sullivan

and pastels by August Will, which are rotated for public viewing. The preeminent New Jersey artist's landscapes depict Jersey City during the second half of the nineteenth century before its urbanization.

—The Committee for the Absorption of Soviet Emigres Museum in Jersey City offers an unusual experience—art not officially sanctioned by the Soviet government. Many pieces were literally smuggled out of the country. Containing work by both Soviet immigrants and artists still in Russia, the 260-piece collection features paintings, sculptures, lithographs, and other graphics.

The northern New Jersey artistic landscape is also amply dotted with privately owned galleries that welcome the public. Some notable collections are Seraphim Fine Arts Gallery in Englewood; American House Gallery in Tenafly; Galleria Maray in Englewood; Aurora Gallery in Closter; Wyckoff Gallery, Inc., in Wyckoff; Kukwa Gallery in Tenafly; Gallery Montage in Midland Park; and Jewel Spiegel Gallery in Englewood. In short, the visual arts are alive in the north.

A fine example of public art, down to earth though ephemeral, is this Little League mural painted on the side of a building in Hoboken. Photo by Carol Kitman

Zimmerli Art Museum is located in Voorhees Hall on the campus of Rutgers University in New Brunswick. Photo by David Greenfield

LITERATURE AND POETRY   New Jersey, on the whole, supports its literati, and the list of writers and poets is impressive. Novelist Joyce Carol Oates, playwright Amiri Baraka, and poets Galway Kinnell, Alicia Ostriker, and Gwendolyn Brooks have all made appearances in northern New Jersey. The annual Dodge Poetry Festival, held last year at the Waterloo Festival for the Arts in Stanhope, offers a splendid forum for these prolific artists.

The Passaic Community College Poetry Center has received numerous accolades for its support of the literary arts. It was named the Distinguished Arts Organization in New Jersey by the New Jersey State Council of the Arts. The organization was also the recipient of a national endowment for its excellent poetry program. Maria Gillan, director of the Passaic County Cultural and Heritage Council, also directs the poetry center. The council has been responsible for financing several outstanding literary projects: *Footwork,* an 80-page collection of poetry, short stories, and book reviews, and the *Mill Street Forward,* an alternative newspaper which has been compared to the *Village Voice* in its early days.

NORTHERN NEW JERSEY ARTS COUNCILS   It would be difficult to overlook the importance of the separate county arts councils in promoting interest and activity in the arts. They are literally the mortar

Expressions of art can be discovered nearly everywhere in northern New Jersey, especially if you look up. The facades of many older buildings are adorned with varied details. This building is located in Hoboken. Photo by Carol Kitman

Every town, even the smallest rural hamlet, is alive with cultural activities like exhibits, readings, theater, concerts, etc. Every taste can be satiated, up to and including the most refined. The region's established cultural institutions—the symphony, the opera, the ballet—are all earning national attention. Even performers from New York and all over the world have made it a stop on scheduled tours. Northern New Jerseyans no longer need to travel to satisfy their aesthetic desires. Just as their jobs now exist within their "neighborhoods," so do their cultural pursuits. In the end, all of this equals an enviable style of life that is sought after by corporations seeking new locations.

No longer consigned to cultural oblivion, northern New Jersey's ascendancy to artistic prominence has been a source of local pride. The same "local pride" that William Carlos Williams fought so hard to establish has come to fruition. According to Mariani, Williams wanted his audience to "see the world under its very feet." Williams did that by celebrating, in an idiom he helped to reshape, the beauty and importance of the things he saw around him:

I have discovered that most of
the beauties of travel are due to
the strange hours we keep to see them:

the domes of the Church of
the Paulist Fathers in Weehawken
against a smoky dawn—the heart stirred—
are beautiful as Saint Peters
approached after years of anticipation.

Photo by Carol Kitman

that holds all the disparate artistic activities together. Without their expertise, coordination of events could not occur. Northern New Jersey is particularly lucky to have active and imaginative county historical, cultural, and arts councils that are set up to respond to inquiries made by the public about upcoming special events.

Over 37,000 professional writers, artists, and entertainers work and live in New Jersey. This year, the State Council on the Arts named 13 northern New Jerseyans as Distinguished Artists. Commenting on the council's support of the arts, executive director Jeffrey A. Kesper said, "With the wealth of talent in this state, the fellowship program has experienced tremendous artistic and financial growth."

Approximately 1,000 arts organizations and agencies also make New Jersey their home, with many situated in the northern regions.

The New Jersey system of higher education has maintained an outstanding record for over 100 years. Photo by E. Lewin

In an age when the nation's economy is in the throes of change from a manufacturing base to information/service, New Jersey's sustained prosperity has made it a model for others to emulate. Its highly educated population (it ranks twelfth in the nation in the number of college graduates) has bolstered that economic success by providing the necessary skilled manpower to support it. Ninety of the nation's 100 largest companies have headquarters and/or operations in New Jersey. These corporations have made "good neighbors," and in many instances have formed unique partnerships with academia, the community, and government. These "symbiotic" relationships have been the basis for enriching every segment of life in the state, and they have helped New Jersey's system of higher education to flourish.

CHAPTER NINE

The great strides the state has taken to promote excellence within its academic communities has drawn national attention and made New Jersey's educational innovations a prototype for the future. Quoted in an article in the *Chronicle of Higher Education*, Ernest L. Boyer, president of the Carnegie Foundation for the Advancement of Teaching and

Education: An Archetype For Excellence

めーめがねーねこ　どもー
ーこがねー　ねずみーみそしる

A future generation of college-bound individuals were found in this Fort Lee classroom. Photo by Bob Krist

chairman of a special commission studying private higher education in New Jersey, calls the state's higher education policies "unusually progressive."

A RICH HISTORY   New Jersey's educational history includes stages of development and growth that occurred in response to societal and economic shifts. From its illustrious start, New Jersey's institutions of higher learning have set an example for others to follow. Its two cornerstone universities were founded before the Revolutionary War, making New Jersey one of the two first states to establish centers of higher education. Princeton, founded in 1746, was America's fourth college. In 1766, Rutgers became the eighth college in colonial America. Ten other colleges in the state are at least 100 years old. Even

Fairleigh Dickinson University, considered "new" by many, is nearly half a century old.

By the nineteenth century, New Jersey was busy creating normal schools (from the French *Normal Ecole*). These provided the training for high school teachers needed in the rapidly growing system of public education. The establishment of normal schools also brought about the first state funding, and the first instance of women as both teachers and students. An example of these institutions' unique mission is clear in a representative paragraph taken from the 1911 catalogue of the nationally known Montclair Normal School (now Montclair State College): "The Montclair State Normal School is a professional school, whose single aim is the preparation of teachers for the elementary schools and kindergartens of the state."

As time passed, New Jersey's normal schools shared the pool of available students with a plethora of colleges run by various religious denominations. The Newark Diocese sponsored the start of Seton Hall in 1856. It became the first Catholic college in America not administered by a religious order. With Catholic education on its way, the

もーもんくーく
るびいーいし

This exceptional museum of geology is part of Rutgers University, New Brunswick. Photo by David Greenfield

Jesuits founded St. Peter's College in Jersey City in 1872. And graduates of St. Elizabeth's, founded in 1903, became the first women to receive four-year degrees. Other denominations also sought to train their youth. Drew University was founded by Methodists in 1866. It was named for Daniel Drew, a Wall Street financier who donated the money, land, and buildings. German Presbyterians started what is now Bloomfield College in 1868, and Swedish Lutherans established Upsala in 1893. At the same time, two technical schools saw their beginnings: Stevens Institute of Technology in Hoboken (1870) and the forerunner of New Jersey Institute of Technology in Newark (1881).

Other schools sprang up throughout the state: Georgian Court, Douglass, Rider, Caldwell, and Felician College in Lodi, a four-year coeducational institution. In 1941 Fairleigh Dickinson University was founded. In fact, it was during this period that higher education in New Jersey met its greatest test when it attempted to meet the educational demands of returning servicemen. By 1941, there were 30 colleges, including six "relief" junior colleges. Approximately 20,000 students took advantage of the system, which was sound but limited in scope.

Private colleges were forced to absorb the impact of returning veterans. The flood of ex-servicemen into the system proved its inadequacy to respond to such numbers. New Jersey's reprehensible lack of state support for higher education became evident in the condition of its state teachers colleges. Although they still offered what most considered the best teacher training in the country, their campuses were run down. In 1951 voters approved a $15-million bond issue, which set New Jersey firmly on its way to rectifying its course and changing its educational landscape. Since then, the state's growing support of higher education has accomplished solid results: normal colleges were made full liberal arts institutions; 19 new community colleges and three new state colleges were created; Rutgers became the state university, and has expanded on three campuses; Newark College of Engineering became one of the largest engineering schools in the country, changing its name to New Jersey Institute of Technology; and the 15-year-old University of Medicine and Dentistry of New Jersey grew to become one of the largest medical colleges in the country. It includes four separate colleges—three medical colleges and a dental college.

The gardens of Cook College of Agriculture at Rutgers University provide a pleasant repose for students and visitors alike. Photo by Robert J. Salgado

IN PURSUIT OF EXCELLENCE The establishment in 1967 of the New Jersey Board of Higher Education and the Department of Higher Education was a further impetus for change. The state embarked on a significant capital and operating investment program designed to make New Jersey's system of higher education the best in the nation. Now celebrating its twentieth anniversary, the New Jersey Board of Higher Education has orchestrated landmark changes, many of which are the result of efforts over the last decade. These commitments have included garnering public support, defining identities or areas of emphasis for their institutions, and setting long-term goals. They are now reaping the fruits of those labors, as New Jersey schools enter the twenty-first century as leaders in their respective fields of expertise.

Commenting on the notable progress, T. Edward Hollander, Chancellor of the Department of Higher Education, notes that "the

system is now complete; what has changed dramatically in the last decade has been our aspirations as a system of higher education…We set our goal to be among the best in every sector of higher education. As a result, we have embarked on major new financing, as well as major new approaches to financing." Explaining the significance of the "new approaches" Hollander said that the state had utilized the business model as a "schematic" to achieve quality:

We borrowed from business and industry by financing many of our innovative programs on a competitive basis. That is, we don't make money available on an entitlement basis—institutions must compete for money beyond the base budget and cost increases. I guess the key is that we've learned a lot from business and industry on how to improve the product.

Pointing to one example of the success of such an approach, Hollander said, "Rutgers is now recognized nationally in a number of its disciplines, and it is beginning to attract Nobel-quality people from around the country. I believe that by the end of the century its standing

Rutgers is the eighth oldest institution of higher education in the United States. The Newhouse Center for Law and Justice is in Newark. Photo by Carol Kitman

The Dahlburg Daisy is one variety of flora found within the gardens at Cook College of Agriculture. Photo by Robert J. Salgado

will be clearly among the top ten and probably even better."

COMPETITIVE GRANT PROGRAMS   Since the early 1980s, following the business community's lead of making financing available on a competitive basis, the Department of Higher Education has awarded approximately $100 million under the Challenge Grant program, which has stimulated institutions to create superior programs in computer science, the humanities, mathematics, science, and other fields. The program has caused considerable interest around the country, and the Carnegie Foundation's Ernest Boyer suggests that such an approach may become "a model for the nation."

Designed to encourage institutions to develop strength by focusing on specific academic areas, the grants have been awarded to the state's three public research institutions, several four-year state colleges, and several two-year county colleges. According to Hollander, not every institution is guaranteed funding: "In every one of those programs, the institution has had to compete for the money. Their proposals have been judged by a peer-group evaluation process using all out-of-state people. It is similar to the way the National Science Foundation

evaluates proposals." As a result of this approach, participating institutions have been able to take a leading role in the excitement of fashioning their own future.

NORTHERN SCHOOLS MEET THE CHALLENGE   With nearly $6 million in challenge funding, New Jersey Institute of Technology (NJIT) has accomplished a great deal over its two-year involvement in the program. One of NJIT's noteworthy contributions involves its key role in advancing the state's Computer Integrated Manufacturing (CIM) centers. The centers will meet current and future needs for CIM technicians in New Jersey. The centers exemplify not only an important curricular concept, but also a new level of cooperation among higher education institutions and between the institutions and the region's business and industrial communities. The Northern/Central CIM Center, located at NJIT, comprises 12 county colleges with NJIT as the lead institution.

NJIT is also proceeding with its plan for a computer-intensive campus environment by adding a new station networked computer laboratory (bringing the total to three). This completes the backbone of the campus-wide fiber optics-based computer network, as well as providing microcomputers for every incoming freshman.

In 1987, the University of Medicine and Dentistry of New Jersey (UMDNJ) received $4.8 million in funds to continue its prodigious advancements in research. UMDNJ has recruited nationally prominent scientists to chair several basic science and clinical departments at New Jersey Medical School and Robert Wood Johnson Medical School in Piscataway. The funds have also supported excellence in the study and treatment of heart and liver disease. The Sammy Davis, Jr., National Liver Institute's research in liver disease, for example, makes it the nation's first medical resource devoted solely to patient care, education, and research for liver disorders. Located on the Newark campus, the center seeks to expand and enhance the work that physicians and scien-

tists at UMDNJ-New Jersey Medical School are doing in the prevention, diagnosis, and treatment of liver disease. In 1985 a national fund-raising drive was launched for the institute with a benefit dinner.

Several state colleges in northern New Jersey took up the gauntlet as well, responding to the challenge in ways that are congruent with their distinctive missions and environmental settings. For example, Jersey City State College sought to blend all its academic programs with a cooperative education experience. Students are required to have an applied learning experience of two six-month semesters in a professional setting relevant to his or her major.

William Patterson College of New Jersey's interaction with the community is the basis for its establishment as an important resource in the economic development of this five-county area. The college proposes to provide a technologically trained work force, conduct applied research, and engage in extensive community service. According to Dennis Santillo, director of college relations, the college plans to become a "nexus of activity that includes the community and the corporate sector." The college is developing centers of excellence in both

Within the state of New Jersey, the government, corporations, community, and academia have combined their interests and formed a powerful team. Caldwell College, pictured here, is a member of that team. Photo by Sharon Sullivan

its science school and communications department. The Center for Applied Science will respond to the need for research, particularly in the fields of biotechnology and environmental waste management. The institution will also focus on educating mathematics and science teachers. In addition, it will set up an Instructional Communications Studies Center that will train telecommunications specialists. Students will be part of a program that stresses management and internships. "It is our intention," Santillo commented, "to prepare men and women for a place in the new economy. What we want to do is to take people from the region into the college, educate them, and send them back to the community as trained professionals and citizens who can contribute."

Ramapo College of New Jersey in Mahwah had one of the most creative responses to the Challenge Grant program. Spurred by the presence of a preponderance of international and multinational firms at its doorstep, the college found that its orientation made it "the college of choice for a global education." The liberal arts college, which is located in New Jersey's most prosperous county—Bergen—is uniquely suited for the task. It will establish a state-of-the-art television commu-

nications center that will be available to corporations for use and will help develop global and multicultural literacy among its students. Ramapo President Robert A. Scott underscores the importance of Ramapo's role in providing such a unique educational environment: "If our students are to become globally literate, they must study the international and multicultural dimensions of their fields in professional programs, general education programs, and even in their extracurricular and civic activities." Scott outlined Ramapo's immediate goals when he stressed that the college would establish

...an institute for American Studies so that the foreign employees of area corporations can study the English language, culture, and history at times convenient to them, through seminars both on campus and at the corporate site; and programs for American employees of foreign firms so that they can learn the language, history, and culture of their employers. Our first objective is to establish such a program with Japanese firms, and we are being assisted in this effort by Sony and Minolta, members of the Ramapo College Industrial Associates.

Ramapo College is a school of liberal arts located in Mahwah. Photo by Vincent T. Marchese

Ramapo College educators participate in a Free Market Conference series of lectures for high school students interested in business related topics. Photo by Vincent T. Marchese

RESEARCH AND DEVELOPMENT   New Jersey's two major research and development areas are impressive examples of activity that draw national attention. Route 1 in Princeton is the home of numerous companies along what has become a 25-mile-long high-tech highway. The longer stretch of I-287 from Edison to Morristown has also attracted many research-oriented companies. The excellence maintained by these giant firms is fortified and supported by the state's many colleges and universities. They have been a source of knowledge and inspiration, and they have provided highly qualified job candidates. In fact, by the mid-1980s, more than 100,000 scientists and engineers were living within the state—the highest per capita in the nation. Northern New Jersey's many institutions of higher education have actively partnered with business to create interesting research projects.

In November 1984, voters approved a $90 million "Jobs, Science, and Technology" bond issue that funded capital projects to strengthen New Jersey colleges and universities in two areas of high technology—research and education/training. To advance this dual purpose, the act specified $57 million for a network of cooperative academic-industrial research centers. These Advanced Technology Centers (ATCs) are administered by the Commission on Science and Technology. One of the most important centers is based in the north in Newark at the New Jersey Institute of Technology (NJIT). The Center for Hazardous and Toxic Substance Management is the nation's first center dedicated to cooperative academic/industrial research in the field. The center is supported by a consortium of four of the state's leading research universities: Stevens Institute of Technology and UMDNJ, both in Newark, Princeton University and Rutgers. The academic partnership is complemented by industrial members who contributed over $600,000 last year to support the center's work.

The Commission on Science and Technology has also established a series of Technology Extension Centers (TEX) in order to help businesses stay competitive by offering access to the latest in technical advancements. Through these centers technical information is trans-

Princeton University was established in 1746 as The College of New Jersey. Photo by Mark E. Gibson

ferred out of the academic laboratory into the workplace. Several institutions in northern New Jersey participate by housing centers.

NJIT is the home of the Information Services TEX Center, which offers assistance to firms needing knowledge in computer technology. The program draws on the expertise of both NJIT and Stevens Institute of Technology.

The Newark campus of the University of Medicine and Dentistry of New Jersey is the home for the Center for Investigational Cancer Treatment. It is under the direction of Dr. David M. Goldenberg, one of the country's leading cancer researchers. Goldenberg, a pioneer in the use of antibodies for cancer detection and treatment, is president of the Center for Molecular Medicine and Immunology at UMDNJ. The center is the core of a statewide network of hospitals involved in the project. In addition, the TEX center is a valuable resource for New Jersey's important pharmaceutical industry because it supports the clini-

cal trials necessary to discover and develop new products.

In 1985, the commission saw the opportunity to utilize the Polymer Processing Institute at Stevens Institute of Technology in Hoboken. The commission now supports the Polymer Extension Center at Stevens. It has been estimated that more than 1,500 companies in New Jersey either supply or process polymers, accounting for more than 100,000 jobs. The center provides industry—particularly smaller businesses—with the complicated technology needed to stay competitive in today's marketplace.

Beyond the work of the Commission on Science and Technology, northern New Jersey's educational institutions have initiated numerous other projects that support research and development efforts in the state.

In October 1987, one multinational corporation announced its gift of $3.6 million in cash, computers, and scientific equipment to benefit higher education. The contribution is one of the largest gifts ever given by a corporation. It will support research and other academic programs at five schools: NJIT and Stevens in northern New Jersey, and Princeton, Rutgers and Monmouth.

In early autumn, the campus of Princeton University is ablaze with color. Photo by Bob Krist

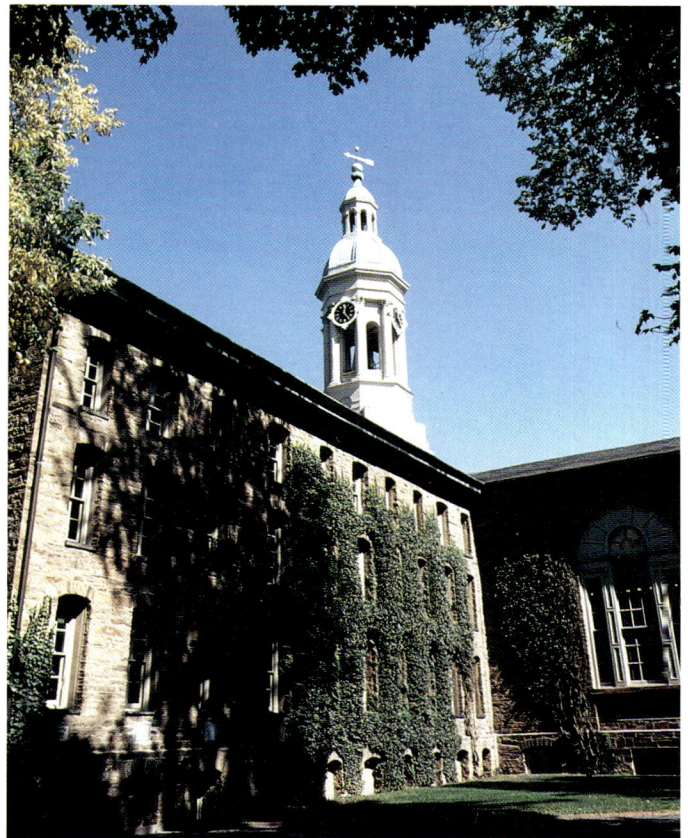

Educational programs offered at Princeton University employ the preceptorial method of teaching: independent reading, small group discussion, and private student-teacher conferences. Photo by Mark E. Gibson

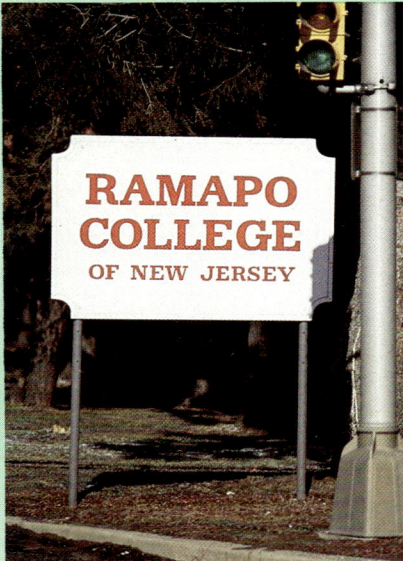

Students have a green-light at Ramapo College. Photo by Sharon Sullivan

This sign points the way to one campus of the largest independent university in New Jersey. Photo by Sharon Sullivan

Drew University is the home of the Charles A. Dana Research Institute for Scientists Emeriti. Photo by Sharon Sullivan

Located on the Teaneck-Hackensack campus of Fairleigh Dickinson University, the brand new George and Phyllis Rothman Center is easily identified by its surrealistic rooftop. Photo by Sharon Sullivan

The Davidson Laboratory at Stevens Institute of Technology is the largest university-based lab for the study of ocean transportation. Models of the world's ocean-going vessels are tested in Olympic-size pools to determine the seaworthiness of the full-size versions. Davidson Laboratory is internationally recognized for its contract research in hydrodynamics, terradynamics, and ocean engineering.

Stevens Institute's commitment to research activities has garnered world recognition. For the last decade, the school has actively encouraged women to pursue technological careers. Under the direction of the Office of Special Programs for Women, a series of career awareness programs was targeted to junior and senior high school women, and their parents, teachers, and counselors. The program prepares young women to become active participants in today's thriving economy. It is supported by area corporations, foundations, and individuals.

Fairleigh Dickinson University operates from three major campuses: Teaneck-Hackensack, Florham-Madison, and Rutherford. At its Rutherford campus, undergraduate chemistry majors have an integrated schedule of classroom studies and on-the-job experience in industry. Industry assignments are matched to the students' specific career interests. In this way, undergraduates obtain a clear view of what their future in research will hold. Dupont, Hoffman-LaRoche, and Eastman Kodak are a few of the participants in the program.

The Charles A. Dana Research Institute for Scientists Emeriti was founded in 1976 at Drew University. The university gives retired scientists an opportunity to continue their research with up-to-date laboratories while affording its students the opportunity to learn from some of the best minds in the corporate scientific community. The institute has helped to strengthen important and mutually beneficial bonds between Drew and its corporate neighbors. The Charles A. Dana Foundation made the institute a reality by a grant awarded to the university. Countless corporate contributions have helped to support it.

THE HUMANITIES   New Jersey's commitment to scientific research has not prevented it from becoming an active advocate of a sound liberal arts education. In the last decade, liberal arts departments around the country have seen enrollments dwindle in traditional liberal arts majors such as literature, languages, philosophy, history, and social sciences. New Jersey has been no exception. However, it is one of the few states taking steps to promote the humanities. The humanities "roundtable" is made up of business and higher education leaders. Its efforts were launched by an advisory committee chaired by Paul Hardin, the president of Drew University in Madison, and Joseph Semrod, president and chief executive officer of United Jersey Banks, centered in Princeton. In an October 12, 1986, interview with the *Star-Ledger,* Semrod said that he believed "the world of business has always required more than technical expertise." The validity of his statement is underscored by studies done in 1984 by AT&T on college experience and managerial performance. The results confirmed earlier conclusions that Humanities/Social Science majors had the best overall records. The report concluded that a liberal arts education better provides individuals with the ability to adapt to a changing work environment.

Such a broad intellectual base helps one rise in any management hierarchy. T. Edward Hollander, the state's higher education chancellor, compared business majors with education majors, contending that both courses of study are too specialized to allow students the opportunity to take courses that would allow them future flexibility in the work place. "In two to three years," he said, "what they've learned is likely to be obsolete." The roundtable will study what kinds of courses are most useful for business careers, and will sponsor events such as on-campus seminars for visiting executives. A long list of corporate leaders and college presidents is represented on the committee. Northern New Jersey's fine liberal arts institutions uphold the tradition to teach what the Victorian writer Matthew Arnold called "the best that has been thought and said" to each generation of undergraduates.

THE FOUNDATION FOR FREE ENTERPRISE   America was founded on the idea of a free enterprise system. Two hundred years ago, our forefathers fought to give that dream substance and reality. Today, the Commerce and Industry Association of New Jersey helps younger Americans to understand the principles of the system upon which the nation was built.

Since its founding in 1975 by a group of energetic New Jersey business executives, the Foundation for Free Enterprise has established a series of unique educational activities for age groups ranging from elementary school students to teachers.

During its first decade of existence, the Foundation, which is an affiliate of the Commerce and Industry Association of New Jersey, concentrated its efforts on an annual summer seminar conducted for high school juniors and seniors. The seminars have traditionally been held at the Sterling Forest Conference Center in Tuxedo, New York. Each year, approximately 100 students are sponsored by area business organizations to attend the four-day conference. By sponsoring students, these business leaders attempt to ensure that students will enter college with an expanded knowledge of the benefits of a free enterprise system. In its early years, the Foundation (as well as professional economists) accepted the responsibility of informing young people of alternatives to government control, asserting that business functioning in a free-market economy is not the enemy of the people, but in fact permits America to grow and expand its position of wealth and leadership among the countries of the world.

Members of the Commerce and Industry Association and the Foundation For Free Enterprise also sponsor other conferences that

Students of Ramapo College learn international and multicultural aspects of their chosen professional fields, thus enabling them to assimi- *late as employees into New Jersey-based foreign companies. Photo by Bob Krist*

foster the principles of free enterprise. In its Young Scholar's Conference, interested students attend a series of 10 evening meetings of two to three hours in length. Each meeting is sponsored by a company or individual, and taught by academicians with specific areas of expertise.

In its Free Market Conference, the Foundation conducts a 10-week course for high school students. Lecturers come from many of the area's finest universities and colleges: New York University, Rutgers University, George Mason University, Pace University, and Ramapo College. During the course, business-related topics are explored in depth. For example, in 1986 the 90-minute sessions addressed topics such as: "How the Market Works," "Monopolies, Cartels and Competition," "Inflation," "Labor and Minimum Wage," "Capital Formation," "Free Trade Vs. Protectionism," "Providing Essential Services: Private Sector or Public Funding," "The Poverty Trap," "Government Intervention and Failure of Socialism," and "The Nature of Capitalism."

In an effort to reach out to the educators who influence American youth, The Foundation developed "Understanding American Business," a course for elementary and secondary educators. More than 1,000 teachers have completed this nationally acclaimed 15-session course of study.

Every year, the Foundation also conducts a one-day workshop for teachers. In 1987, the program was entitled, "The American Economy—Confronting the Issues." Conducted in cooperation with Becton Dickinson and Co., the pharmaceutical manufacturing firm, the workshop addressed various topics of concern: "Taxes, Deficits and Government Spending," "American Industry in a Global Economy," "Privatization—A Viable Alternative," and "Financial Institu-

tions and Investment Vehicles." The programs are offered to teachers at no cost.

Through the Foundation's efforts, a chair of business enterprise was established at Ramapo College to accommodate dissemination of knowledge and experience by New Jersey's top business and industrial leaders.

In its newest venture, the Foundation has created a program for elementary school children. With courses specifically tailored for the youngster, the three-day program introduces fifth graders to economic concepts. In its pilot program, 82 students learned the intricate economic machinations of manufacturing and selling automobiles. The students were divided into "companies," each charged with developing the marketing concept for its car. Volunteer lecturers and professionals in auto sales and marketing were present with advice and guidance. Students making presentations were awarded plaques.

By conducting courses suitable to reach every age and educational level, the Commerce and Industry Association of New Jersey and its affiliate, the Foundation For Free Enterprise, hope to inculcate the principles of free enterprise, which have made and still make the United States a leader among nations.

THE ACADEMIC DECATHLON   Northern New Jersey's business and academic communities are responsible for initiating one of the most adventurous projects to foster academic achievement in the state. Borrowing its name from the grueling athletic contest, the academic decathlon is equally challenging, with events in subjects that range from economics to essay writing.

Using the national contest as its model, businesspeople and academicians in the north formed the Academic Decathlon of New Jersey as a non-profit corporation in 1985. Its progenitors comprised its first board of trustees: B. Franklin Reinauer II, former chairman of Reinauer Petroleum Company, was elected president; Harold B. Conant, president of Health Care Recruiters, was vice president of business; Dr. Roy Stern, superintendent of schools in Cresskill, was vice-

The state of New Jersey requires that every student entering a university system at the freshman level be tested for fluency in all subjects. Individuals are not allowed to pursue a course of study until they are deemed competent. Photo by Vincent T. Marchese

Newark's Free Public Library was once the working place of novelist Phillip Roth. Roth is well known for such works as **Portnoy's Complaint** *and* **Goodbye, Columbus.** *Photo by Carol Kitman*

president of academics; Kenneth Cadematori, managing partner of Price-Waterhouse, was treasurer; and James Cowen, president of the Commerce and Industry Association, was elected secretary.

Business leaders on the board worked at recruiting schools and raising funds to support the decathlon in an effort to spur interest in the events. As a result of their efforts, 31 high school teams participated in the first year's roster of events. The winner, Region High School of Old

of students to achieve excellence. According to Reinauer, the decathlon serves to inspire students, "and it assures us that our country's citizens, workers, employers, employees, and business and government leaders are the best we can produce."

SOMETHING FOR EVERYONE There are no obstacles to attaining a college education in New Jersey. Everyone has an equal opportunity, regardless of his station in life, financial status, or needs. County community colleges are open-admission. Anyone who is 18 years old or who holds a high school degree or General Development Certificate (GED) is guaranteed admission. If a person graduates from a community college with an Associate degree, he or she is guaranteed admission to one of New Jersey's state colleges. According to Dennis Santillo of William Patterson College of New Jersey in Wayne, "Anyone who is serious about getting an education can get one in New Jersey. We encourage everyone because we know that an information society requires an educated work force."

Financial considerations do not pose a problem, either. The relatively low costs of New Jersey's institutions of higher learning bring it within the realm of possibility for most people. Even the most expensive colleges give students financial assistance. In fact, over one-half of all state college students are presently receiving some form of financial aid.

AN EDUCATIONAL CENTER Northern New Jersey has every type of college. Its two-year colleges—primarily county community colleges—exist in abundance alongside four-year colleges and universities, which can be public or private institutions. Post-secondary schools also abound. Most of these do not award degrees. They include the School of Health-Related Professions at UMDNJ, hospital schools, area technical schools, and proprietary schools like the Berkeley School and Katharine Gibbs School.

The cluster of schools in northern New Jersey's urban and suburban areas makes it an educational center. Every kind of professional school, as well as other institutions of higher learning, are easily found there. The preponderance of schools makes any town only a few minutes' drive from a quality educational experience. For example, Bergen County has several colleges and a university within its borders. Each offers its unique qualities. Fairleigh Dickinson University, which opened its doors in 1941, has two of its three campuses in Bergen: Teaneck-Hackensack and Rutherford. Its colleges include Ramapo in Mahwah, the only four-year state school. Felician College in Lodi specializes in a liberal arts education for women. Its continuing education offers a variety of programs for men and women seeking part-time study for professional and personal enrichment. Bergen Community College in Paramus offers a two-year education in career courses, and adult and continuing education. Edward Williams College in Hackensack provides educational opportunities for those who prefer a private junior college approach. These institutions provide a chance for everyone to have access to good educational facilities.

LIBRARIES America's second library was established in New Jersey in 1750 by Dr. Thomas Cadwallader, a friend of Benjamin Franklin. It is no wonder that New Jersey today has a splendid network of over 4,000 public, school, college, and university depositories.

The region's public library system brings local libraries into

Tappan, went on to place thirteenth in the nation. Kathryn Johnson, one of the team members, won a gold medal for her essay.

By its second year, 59 schools participated, and the geographic area had expanded considerably. At this writing, 93 schools from 11 counties have agreed to participate in four regions. Due to all these efforts, the state now ranks ninth in the nation.

The Academic Decathlon provides the motivation for thousands

The state of New Jersey places great emphasis on minority programs, counseling, and scholarships, ultimately aiming to halt the decline in minority enrollment. Photo by Bob Krist

"area" library networks. If a request for data or material cannot be met, it is forwarded to the area library. If necessary, the area library calls upon the major centers of the New Jersey State Library at Rutgers and Princeton, and the Newark public libraries. A network of special libraries gives industry access to information in the chemical, pharmaceutical, telecommunications, and other fields that have heavy representation in the state. Corporate and academic libraries make almost any business and industrial subject easy to research.

A PARADIGM FOR THE FUTURE   New Jersey's aggressive strides toward excellence in education have catapulted its institutions, over the last few decades, from relative obscurity to national attention. Its landmark innovations have been emulated by others who seek to solve some of the more tenacious problems that beset American education in the 1980s.

New Jersey's "Alternate Route" is solving the problem of teacher shortages by allowing qualified people with traditional baccalaureate degrees the opportunity to enter the teaching profession. New Jersey is the first state in the nation to attempt to do this, although others are now seeing the wisdom of this approach. New Jersey is among the first states to take steps to "debureaucratize" its system of state colleges. By July 1, 1989, the transition toward autonomy for state colleges should be complete, and responsibility for managing the nine state colleges will be in the hands of the colleges' boards of trustees.

New Jersey is also the only state in the country to test every freshman upon entrance. On the basis of that test every college is required to provide remedial education, if necessary; it is not optional. Students are not allowed to take courses until they have demonstrated competency in the area in question. New Jersey's emphasis on minority programs seeks to halt the decline in minority enrollments, which many other states are also experiencing. Projects underway offer pre-college counseling, minority scholarships, and contact with college campuses as a means toward achieving better participation by minorities.

Although it has been in the vanguard of so many modern improvements to education, New Jersey's greatest contribution lies in its ability to create lasting and mutually beneficial relationships among its many business and industrial "neighbors," the community, and academia. The state seems to have found the "formula" for productive and happy coexistence. According to Hollander, "Business in New Jersey is very public spirited and they are anxious to be part of the communities in which they 'live.'" He adds that

there is no reluctance here among academics to work with business and industry. We don't see a strong separation between the theoretical academy and the practical world; we think the two are closely tied together. Business, here, is seen as kind of benevolent, wanting to support but not dominate or intrude.

These symbiotic relationships are especially evident in northern New Jersey, where so many corporations are located. The abundance and proximity of the area's corporate "residents" has fostered interest in partnerships that will be of benefit to the region's future. Their success could be used as a prototype for excellence and productivity.

The Jersey City Public Library is just one participating branch in a large network of New Jersey depositories. Photo by Carol Kitman

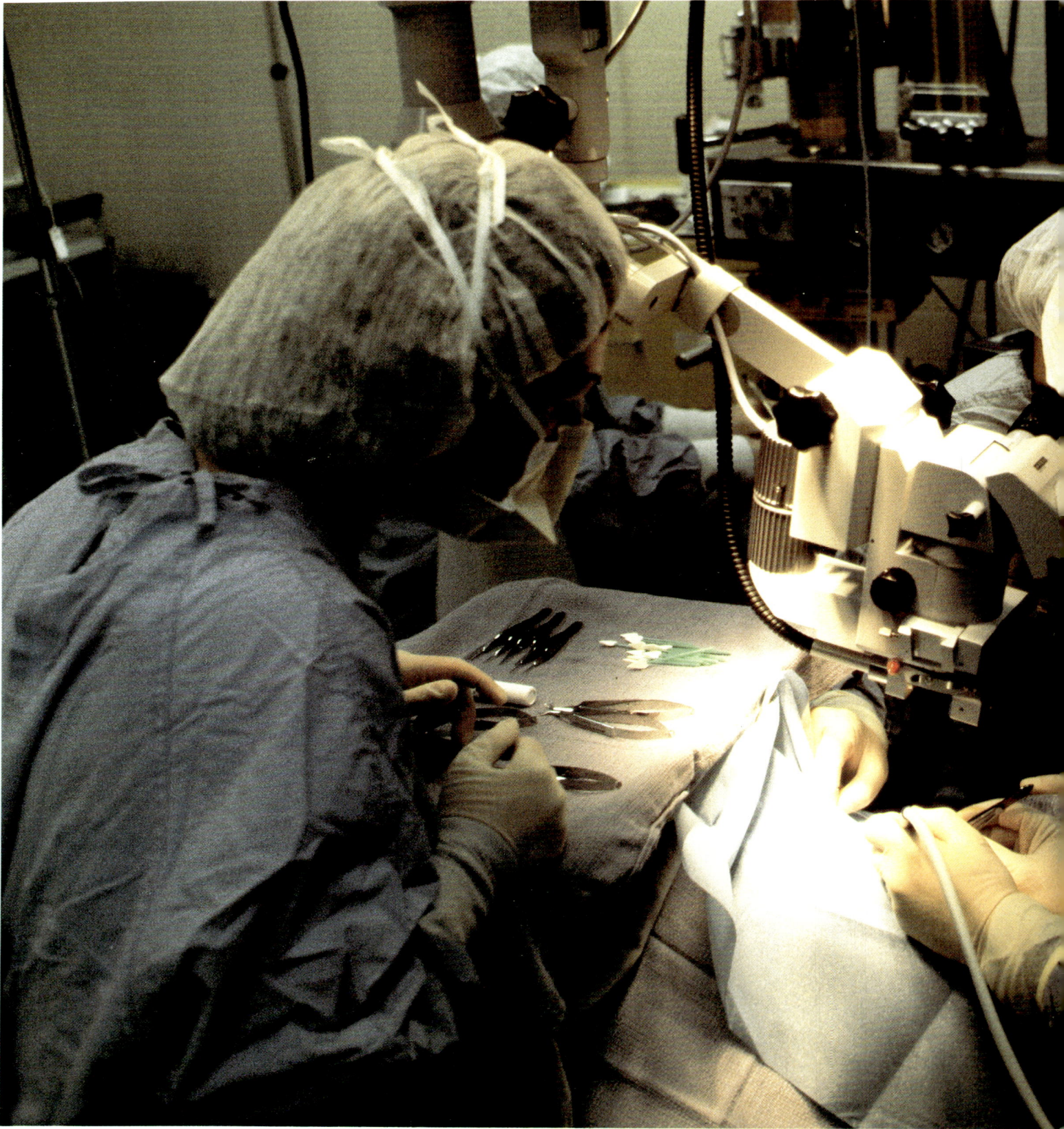

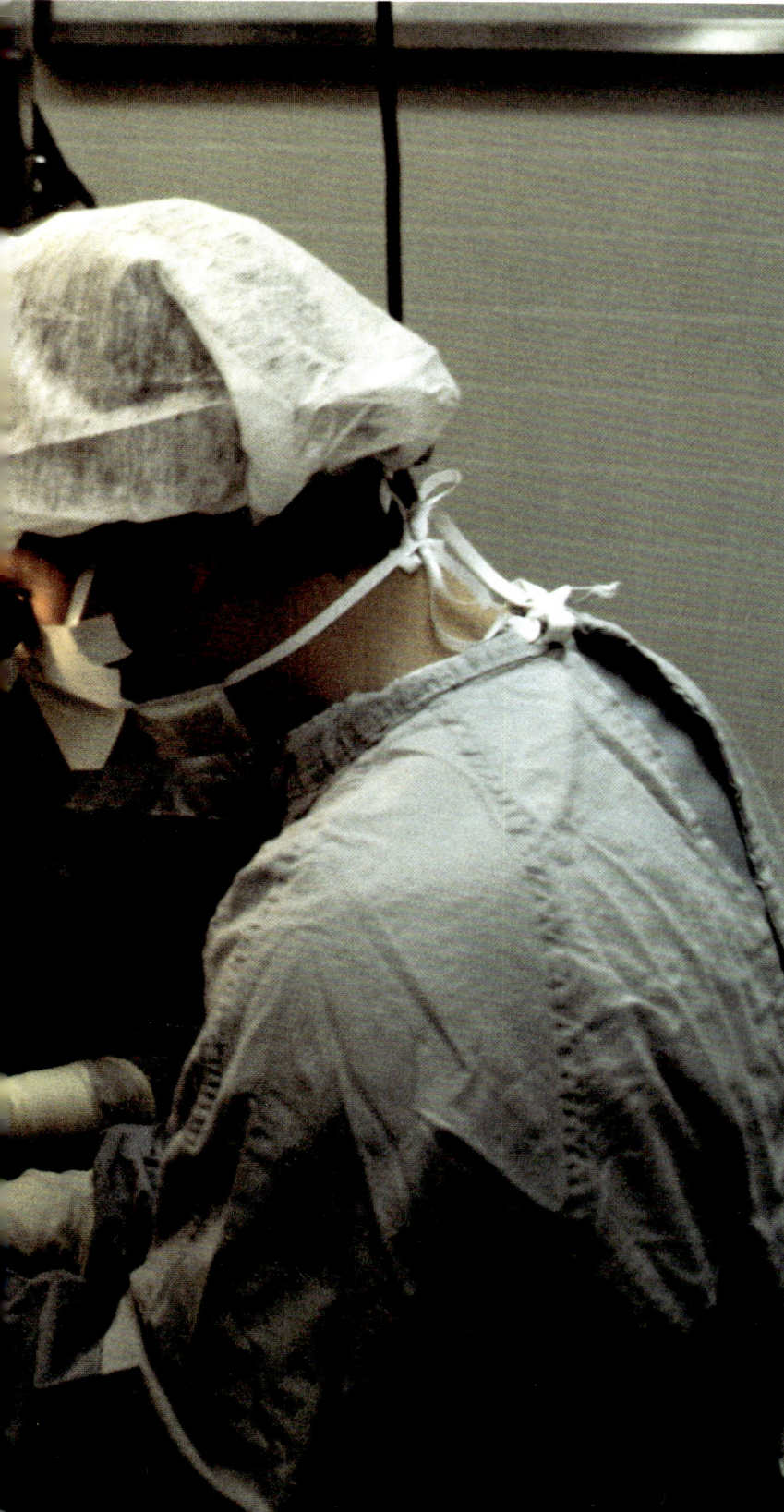

Today microsurgery is performed either by micromanipulator, or laser beam. Photo by David Greenfield

At a time when health care costs are soaring, a national study conducted by the Equitable Group Health Insurance Company cited New Jersey's hospitals as the nation's most cost effective, with the lowest charge for care in the country. According to the data released, there were wide variations in hospital costs nationwide. For example, the average daily charge for hospital care in Nevada, which is the highest, is $915 in comparison to New Jersey's low figure of $437. Likewise, New Jersey's charge for an average hospital stay of 7.4 days is $3,234, compared to a national average charge of $3,840 for an average stay of 6.4 days. Despite careening health care costs across the nation, New Jersey's hospitals have maintained high standards, successfully juggling low cost and quality care.

CHAPTER TEN

According to the New Jersey Hospital Association (NJHA), a proactive advocate for the state's hospitals, this impressive record of cost containment can be attributed to efficient hospital management, the state Diagnosis Related Groups (DRG) reimbursement system, a rapid transition to outpatient and ambulatory services, better home health care delivery, and technological advances. In an article in *New Jersey Business,* Ronald J. Czajkowski, director of press relations for

Health Care: In The Vanguard

These nursing students, with the guidance of the staff dietician, learn to weigh a newly admitted patient correctly. Photo by David Greenfield

NJHA, stated that "New Jersey has the best health bargain in its hospitals; cost containment efforts are strictly enforced and have been for the past six years—sister states have been doing this for only about two."

In fact, New Jersey has been a pacesetter. Its DRG reimbursement system has served as the model upon which the United States government based its payment system for Medicare patients. By 1983, DRGs were written into federal law as the funding mechanism of Medicare. According to a spokesman for the Health Research and Educational Trust of New Jersey (HRET), Washington was under pressure to develop an alternative reimbursement system to save Medicare dollars, and New Jersey's approach was the most promising and innovative at hand.

New Jersey's DRGs, implemented in 1980 and inclusive of all patients, have forced hospitals to be efficient. Previous to their implementation, most payers in this country had paid hospitals the "usual, customary, and reasonable" charges for whatever treatment the hospital rendered. The system actually rewarded hospitals for increased costs. Since hospitals were paid for the care they actually delivered, they made more money by delivering more services. In an attempt to solve the problem in New Jersey, the Department of Health set payments in advance for the type of illness treated rather than the days spent in the hospital. Put in simple terms, DRGs classify all illness into 468 categories or diagnostic groups. A single price is set for each DRG based on the average bill for patients in that DRG across roughly similar hospitals. Hospitals, then, receive only a fixed price per patient set by the DRG appropriate to the diagnosis. Payment is not affected by the services actually provided or what they cost the hospital. According to NJHA, as a result, admissions are down, lengths of stay are down, and acute-care hospital beds are empty. Technology and pharmacology have improved, and procedures once requiring a few days in the hospital are now done on an outpatient basis.

New Jersey's creative steps to curtail skyrocketing health care costs, however, signify only one area in which it is setting standards. The state has also set its sights on ways to solve the nursing shortage problem, which plagues all American hospitals, and it has already found a humane and constructive way in which to deal with the indigent care problem. Its health care executives are meeting challenges to appeal to managed-care insurance plans by consolidating forces and diversifying services. Corporate restructuring is allowing many facilities to venture into a variety of for-profit business activities designed to render support to the main acute care hospitals. In every area, New Jersey's hospitals have been impressive in their response to the demands put upon them by what *Newsweek* called an industry beset by "perennial tumult." In *The Revolution*, Gregg Easterbrook's examination of modern medicine, he maintains that "only in nostalgic reveries were there ever 'good old days' when the norms of medical practice stood still."

A CENTURY OF CHANGE In order to appreciate the vitality and inventiveness of New Jersey's response to the rapid changes in health care, it helps to realize fully the enormity of those changes over the last century. The evolution of the modern hospital has not been a slow one. No American social institution has undergone such a complete transformation.

During the eighteenth century, hospitals housed the hopeless. These houses of "despair and disease" were established to cater to the sick poor and transients. They did little more than provide rudimentary

Patients participating in a cardiac rehabilitation program are monitored constantly by hospital staff. Photo by David Greenfield

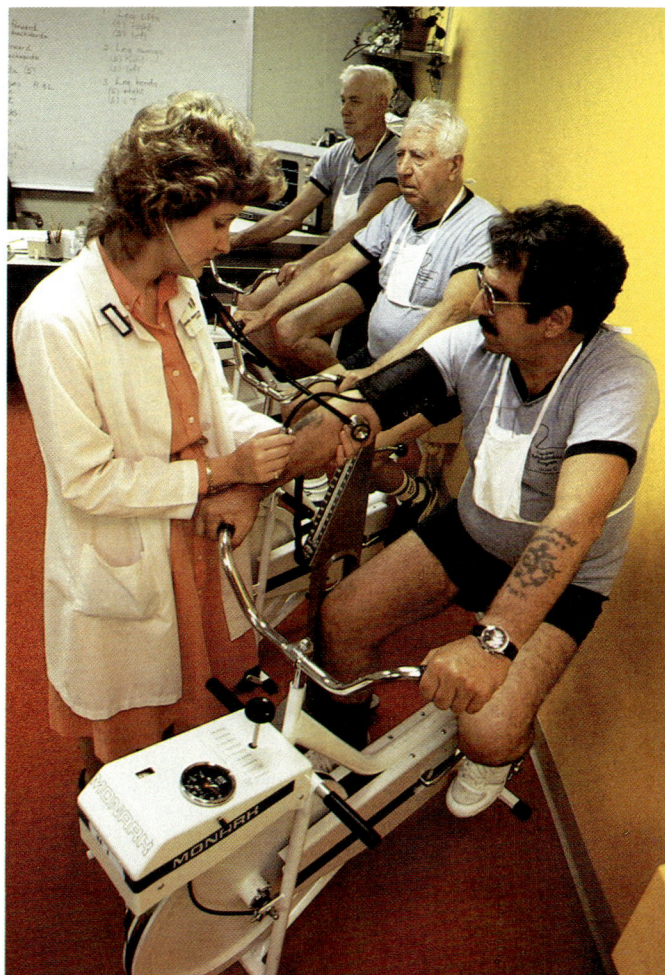

*Fulfilling the demands of contempo-
rary society, many in-hospital alcohol and drug abuse units have been
developed. Photo by David
Greenfield*

nursing care. It was not until the late nineteenth century that hospitals
began to benefit patients. Until then, infection, dirt, and disorder made
hospitals notorious repositories for disease and contagion. "Hospitalism," or cross-infection, threatened all patients.

Louis Pasteur and Joseph Lister both greatly contributed to the
advancement of antiseptic surgery, and that diminished the incidence
of hospitalism. But it was Florence Nightingale's crusade to remake the
hospital that provided the international leadership to change things.
Her experiences in the Crimean War demonstrated to her that
improved nursing practices could reduce mortality. She wrote: "The
very first requirement in a hospital is that it should do the sick no
harm." Her arguments were noticed almost immediately in the United
States by W. Gill Wylie, the most enthusiastic advocate of her doc-
trines. By the century's end, Henry C. Burdett continued Nightingale's
fight by emphasizing planning and rationalization of the hospital's
internal order. His arguments drew international attention. The American Hospital Association came into being by the turn of the century,
and the movement to create efficient, accredited hospitals gained
momentum. These efforts transformed hospitals from medical mast-
odons to the facilities that bear a closer resemblance to our present-day
institutions.

When W.C. Röntgen described his "new kind of rays" in 1895,

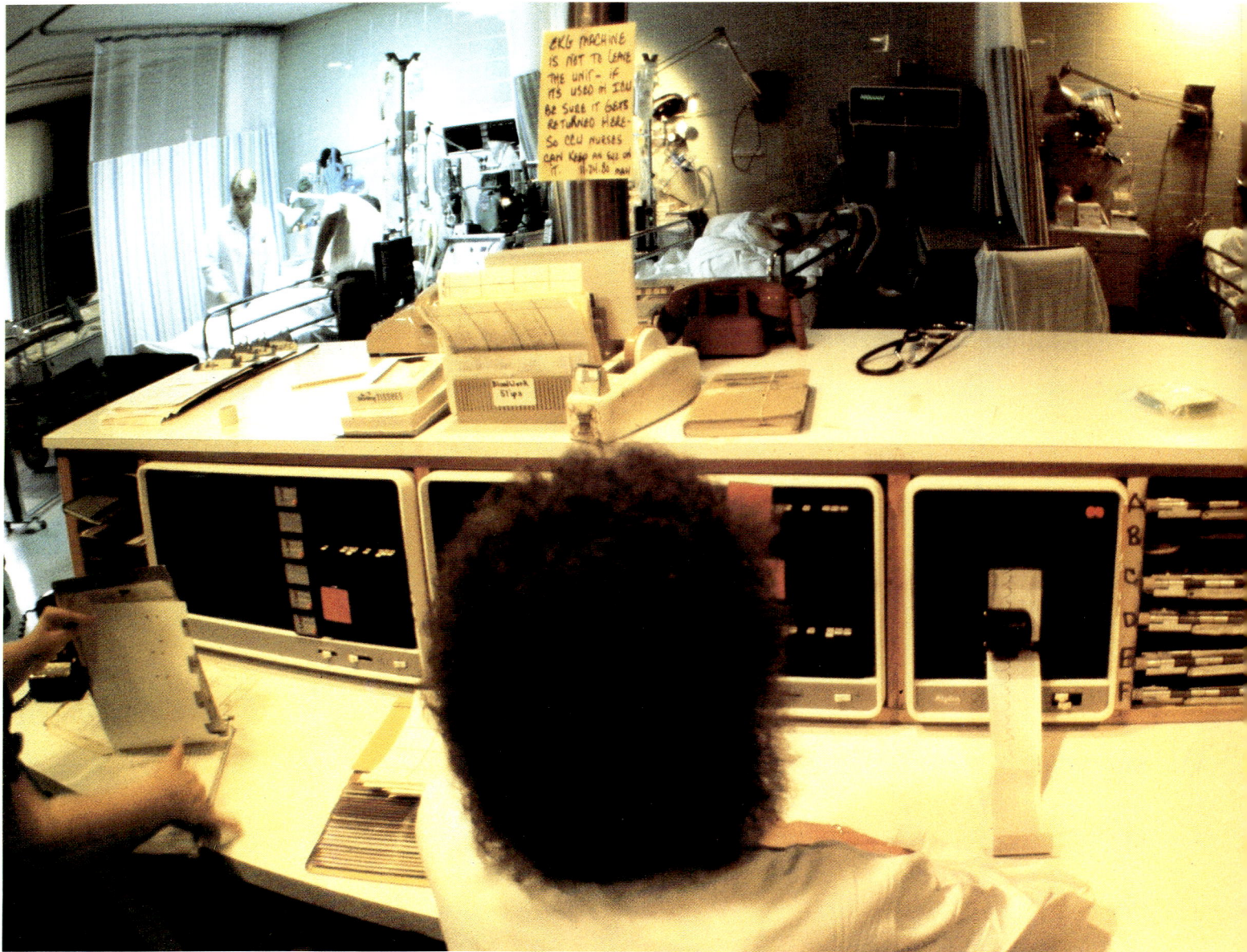

Coronary care units for cardiac pa-
tients have been in use since the mid-
1960s. Photo by David Greenfield

the modern technology of radiology (or roentgenology) and radiotherapy was born. By the 1900s, most major urban hospitals in the United
States had x-ray machines. By 1915, most had separate departments of
radiology. By the 1930s, barium salts and a wide variety of radio-
opaque substances made it possible to "visualize" almost all the body's
organ systems. The new technology made accurate diagnosis of a great
many diseases easier, and gave further impetus to the metamorphosis
of the hospital from passive custodial havens for transients and those
who were sick and poor to institutions taking an active, curative role in
medical care. The introduction of sulfanilamide in the mid-1930s and
penicillin in the early 1940s further helped to suppress infection suffi-
ciently to allow surgery without prohibitive mortality.

The early part of the twentieth century was also marked by controversy over doctors' credentials. The now famous Flexner report con-
demned the inferior quality of medical education in this country. As a
result, as pointed out by the *Newsweek* article, "bogus medical schools
were closed, standards became more stringent and an overall goal of
'scientific medicine' was formulated …American doctors would become mainly scientists, right down to their white lab coats."

In the two decades following World War II, hospitals became the
central repository for the rapidly expanding medical technology. The
growing population turned more often to physicians for help, and
physicians in turn utilized the hospitals' technology more frequently.
As a result, both the patient and doctor were drawn closer to the hospital. During these years, in addition to the creation of high technology, vaccines were discovered which eradicated the dreaded effects of
polio and measles. In light of these significant and visible advancements, the American public began to demand high-quality care and
cures as a matter of course. The images of both the doctor and hospital
were altered forever.

A computer facility is vital to both the suprastructure and infrastructure of the hospital. Photo by David Greenfield

The Robert Wood Johnson Hospital Emergency Entrance is utilized by an ambulance. Photo by Sharon Sullivan

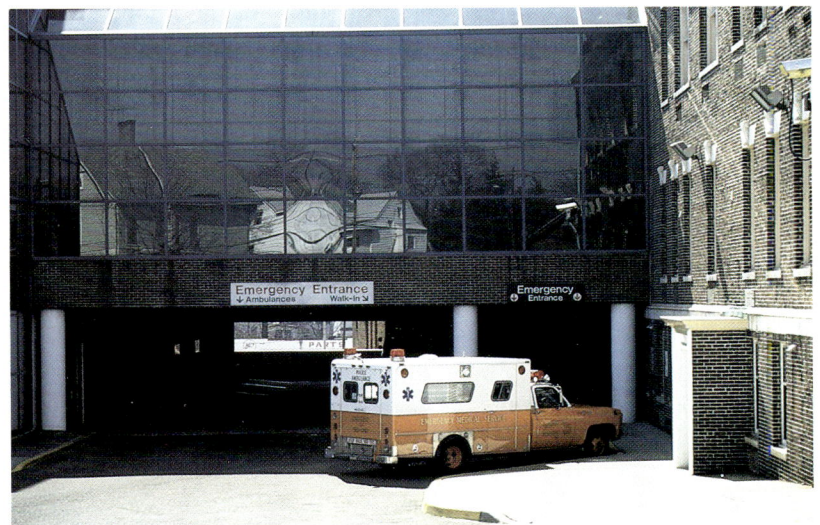

The 1970s saw the advent of intensive-care units as standard facilities in most hospitals. The establishment of trauma centers— which were staffed round-the-clock by surgeons—gave victims of accidents and violence a greater chance of survival. By this time, hospitals were applying the latest technology and procedures to extend the lives of the terminally ill. More and more people were living longer, many sustained only by machinery. Needless to say, the cost of sustaining such a system grew commensurately with the system's increasing complexity. In an article on hospital costs, John Knowles gives evidence of the steady leaps taken by health care costs:

In 1925, the cost of one day's hospitalization at Massachusetts General Hospital was $3. The bill was paid directly by the patient out of his own pocket, and he stayed for 15 days for a total cost of $56.20. In 1972, a patient would stay for an average of seven days at a total cost of $1,400, and the bill would be paid by any one of a variety of third parties.

Knowles points out that between 1962 and 1972, the percentage of hospital income from direct patient payments declined from 38 percent to 12 percent. In the same time period, income from Blue Cross and state welfare departments increased, and Medicare (instituted in 1966) assumed 27 percent of hospital income. By 1982, total United States health care spending exceeded 10 percent of the gross national product for the first time. New Jersey's Diagnosis Related Groups provided a means for putting a cap on the runaway expenditures.

In less than 100 years, health care has metamorphosed completely. However, the torrent of change has not abated. In more recent years, hospitals have had to cope in an increasingly competitive market. Physician advertising, along with surgical clinics and free-standing diagnostic and treatment centers, has reshaped the medical "landscape." Managed Care insurance plans are replacing traditional indemnity insurance coverage. A managed care contract means subscribers will utilize a particular institution. The hospitals and doctors caring for patients in the most efficient ways win contracts with health

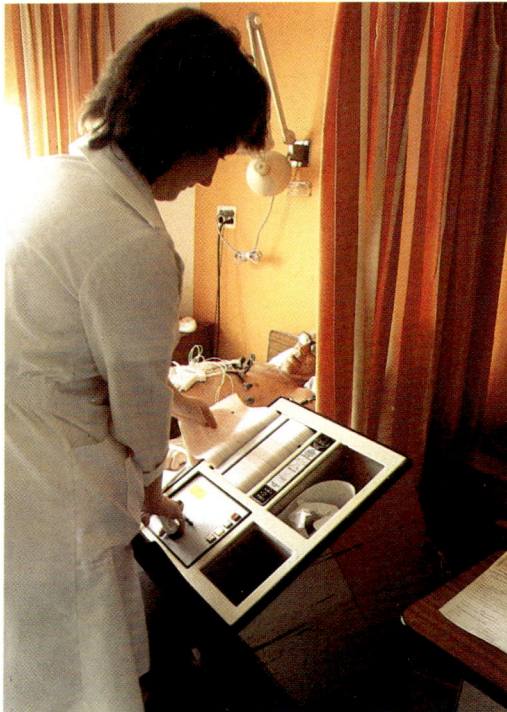

An electrocardiograph records the changes of electrical potential which occur during the heartbeat and is used in diagnosing abnormalities. Photo by David Greenfield

Robert Wood Johnson University Hospital is one of three core teaching hospitals in northern New Jersey. Courtesy, Robert Wood Johnson University Hospital

maintenance organizations (HMOs), preferred provider organizations (PPOs), and similar providers.

New Jersey has demonstrated a capacity to cope successfully and transcend obstacles in the health care environment. The state's health care institutions have managed to retain quality while keeping their costs the lowest in the nation. The industry's most recent attempts to remain flexible and responsive to current needs and trends have kept it in the forefront as an innovator in the field.

CARING FOR THE INDIGENT   New Jersey's hospital industry reasserted its leadership as an innovator in health care policy when it solved the problem of indigent care, which faces hospitals across the nation. While some states practice "dumping," or transferring indigent patients to other hospitals, New Jersey hospitals have a proud tradition of caring for their poor.

In January 1987, state hospitals extended that tradition by pledging to abide by the Uncompensated Care Trust Fund Act. The law

guarantees that no one in the state will be denied necessary care due to an inability to pay. It also protects hospitals in lower-income areas from financial insolvency due to the high volume of care they provide to the indigent.

Calling it a "constructive solution" to indigent care, New Jersey Hospital Association's Czajkowski points out that the trust fund preserves the integrity of the all-payer system, in which all payers are guaranteed access to care. "At the same time," he says, "it adds on an equal flat rate of 10 percent to all hospital bills statewide. That 10 percent is dedicated to uncompensated care." The trust fund eliminates inequity by creating a pool. Hospitals with larger uncompensated care loads receive the difference from the pool; those with lower uncompensated care costs submit the difference to the fund. "In effect," Czajkowski says,

the more affluent hospitals are helping to subsidize the uncompensated care of inner city facilities. This may not seem fair, but if something

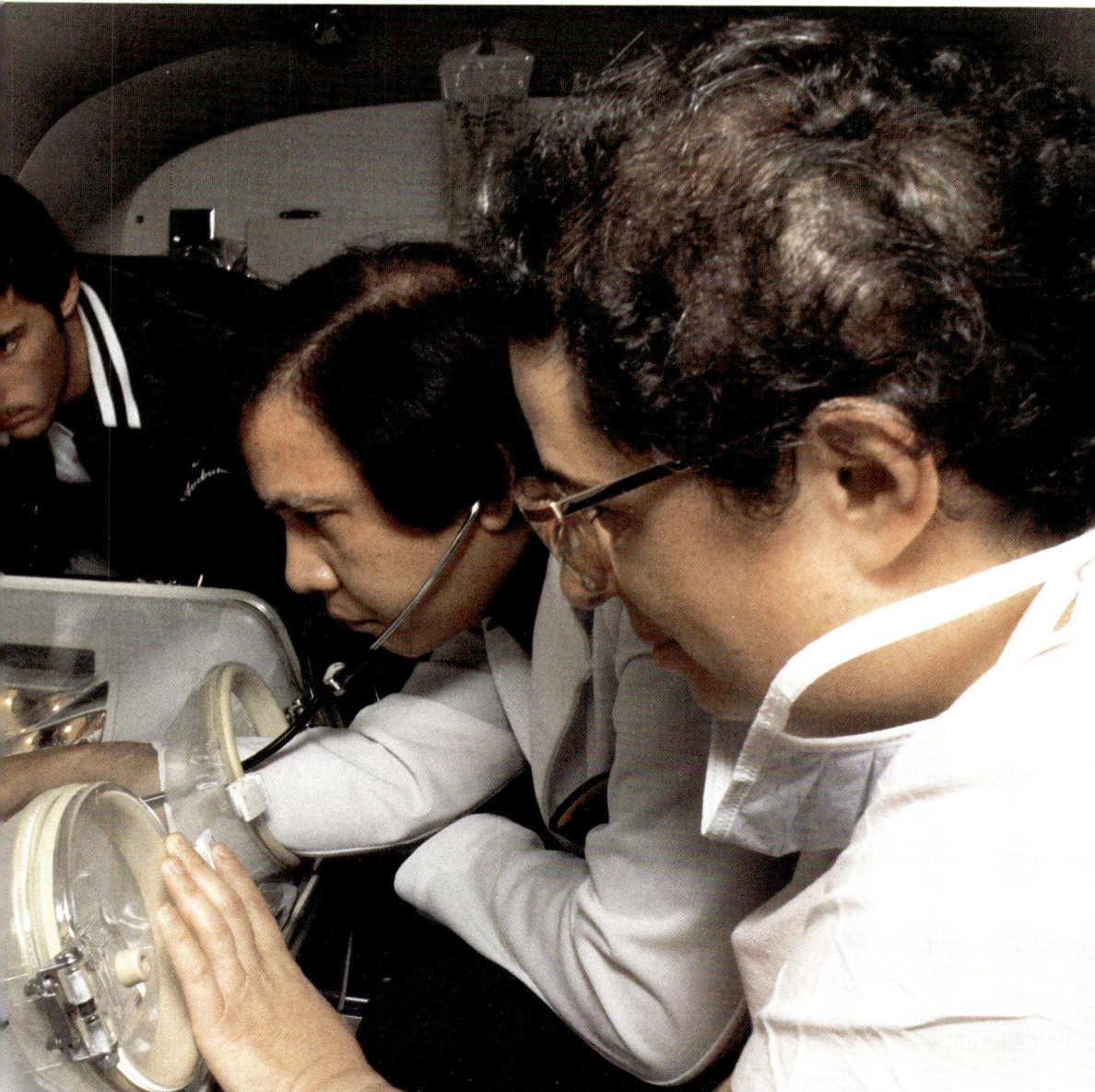

An ambulance team transports a premature infant to a new facility.
Photo by David Greenfield

weren't done to equalize the impact, those hospitals burdened with large numbers of uncompensated care patients would become insolvent. The hospitals left would have to absorb the load.

Czajkowski points out that a few other states have tried to enforce mechanisms that would address indigent care problems; however, New Jersey's fund is unique. According to Czajkowski, the trust fund's two-year "sunset" provision will allow the state to plan "some long-term solutions that people are working on right now."

Last year, New Jersey had $375 million in uncompensated care. According to Louis P. Scibetta, president of NJHA, "the measure combines compassion with practicality by creating a trust fund that ensures the poor receive access to health care, and that the cost of that uncompensated care is spread among all hospitals." Commenting on the new law in an article in the year-end issue of the *Star Ledger,* Health Commissioner Dr. Molly Joel Coye said, "it is an exciting development and one which sends a message from New Jersey that much of the

country is missing. With hospitals in other states turning uninsured patients away, New Jersey is demonstrating that it is possible to care for the poor and still control the rate of increase of hospital costs."

THE "GERIATRIC IMPERATIVE"   As the population of the United States grows older, the concern for adequate long-term care for the elderly increases. In New Jersey that concern is being translated into action as the state's institutions mobilize to meet the demands of the increasing numbers of elderly. New Jersey has the second oldest median-aged population in the United States, behind only Florida. The number of elderly people in the state has increased 32 percent from 1970 to 1985. It is projected that by the year 2000, almost 15 percent of the state's population will be over 65. In an article by Anne Somers, an adjunct professor at the University of Medicine and Dentistry of New Jersey, Robert Wood Johnson Medical School, "New Jersey hospitals are adapting their services and facilities to meet the changing needs of an increasingly elderly population."

The nursing profession averages a 13.6 percent position vacancy rate annually. Photo by David Greenfield

Clearly, New Jersey's hospitals are not ignoring the elderly. In response to the need, they have developed home health programs, provided case management, established satellite geriatric clinics, entered the nursing home business, set up day care for adults, and sought a wide range of alternatives to institutionalization. Somers cites Morristown Memorial Hospital as an example of activity occurring throughout the state. The hospital's Center for Geriatric Care began providing community-based services in January 1985, which includes comprehensive assessment, education, case management, and respite care.

A 1986 report compiled by the Health Research and Educational Trust of New Jersey substantiated the progress New Jersey hospitals had already made toward responding to demographic demands by developing new and innovative ways to deliver long-term care. The report outlined several obstacles impeding progress, the most important being financial resources. Eighty-two percent of responding hospitals cited the lack of "reimbursement" as a major barrier to initiation of long-term care (LTC) services. According to Somers, Medicare's reluctance to reimburse for LTC "either in nursing homes or at home, appears even firmer than a decade ago." Meeting the health care needs of the aged population can be costly, but New Jersey's institutions are overcoming obstacles, with activity in every sector of the state. For example, Alzheimer's disease patients are getting some attention. New

Jersey has approved funding for six geriatric centers in teaching hospitals and is now considering an $11-million program for community-based home care.

THE NURSING SHORTAGE According to the latest figures, the national vacancy rate for registered nurse positions is 13.6 percent. Each year concern mounts as hospitals try to get the professional help they need to maintain quality care. In New Jersey the full-time registered nurse vacancy rate is 17 percent. The situation has been exacerbated by the loss of nurses to adjacent, out-of-state urban-area hospitals, which offer the lure of higher salaries. New Jersey nurses receive an average pay of between $12 and $19 per hour, while those in New York receive over $13.44 to start, and $15.04 in Philadelphia.

According to surveys compiled by JNHA, nurses are leaving due to low pay, irregular hours, outside job opportunities, and the increased burdens that acutely ill patients place upon them. Those who remain in the profession must cope with longer, more stressful shifts due to less staffing. In an aggressive effort to put a halt to any further losses, New Jersey hospitals have been seeking ways to increase career incentives and to alter the public perception of the profession.

Czajkowski notes that NJHA's efforts have constituted solid steps to addressing the problem: "We have been doing the juggling act very well by containing costs while offering access to quality care, and, at the same time, maintaining workable staffing levels...We have formed a nursing shortage task force in order to find a plan to draw more people into the profession." The task force, which is broken into four

Cardiac stress tests include a variety of activities. Photo by David Greenfield

subcommittees, includes representatives from the State Board of Nursing, the State Nurse's Association, nurse educator and executive groups, and the state departments of health, human services, and higher education. The association hopes that the plan will be utilized as a guideline for the industry when it is completed.

AN EMPHASIS ON COOPERATION   Keenly aware of the competitive health care environment in this country, New Jersey's hospitals are doing something about it. By emphasizing specialty areas, hospitals have become centers for excellence. "The acute care hospital can no longer be all things to all people because of the competitive and regulatory environments," Czajkowski commented in a recent interview. He continued:

They provide the basic elements of acute care, but beyond that they don't have the manpower nor the capitol to get involved in every type of service a region or community may need. What they are doing is targeting key areas or strong-suit areas, so that they retain a broad-base reputation while adding a specialty area of expertise.

As a result, each hospital has developed a considerable reputation in one or another specialization. Hospitals have become highly defined

by acquiring specificity, and, in many cases, rival urban medical facilities.

Regionalization is another factor contributing to reducing competition. Consolidations and affiliations have turned previously fierce regional competitors into allies. Facilities have found that they can better keep pace with medical advances by sharing the cost and use of expensive, high-tech equipment. At present, over 15 hospitals in the state are involved in varying degrees of intimacy, including joint ventures, consolidations, and mergers. According to Czajkowski, "a true merger is seen with St Clares and Riverside hospitals in Morris county." In an article in *New Jersey Business*, Czajkowski described the two categories:

A consolidation is seen with Morristown Memorial and Overlook Hospital in Summit. Their board structures are consolidated to reach common decisions regarding joint venture projects such as pooling monies in purchasing a major piece of equipment. A joint venture is a looser approach to affiliation. For instance, four hospitals in a forty-mile area might want to build a hospital-owned nursing home.

Operating room procedures and standards have become more and more stringent. Photo by David Greenfield

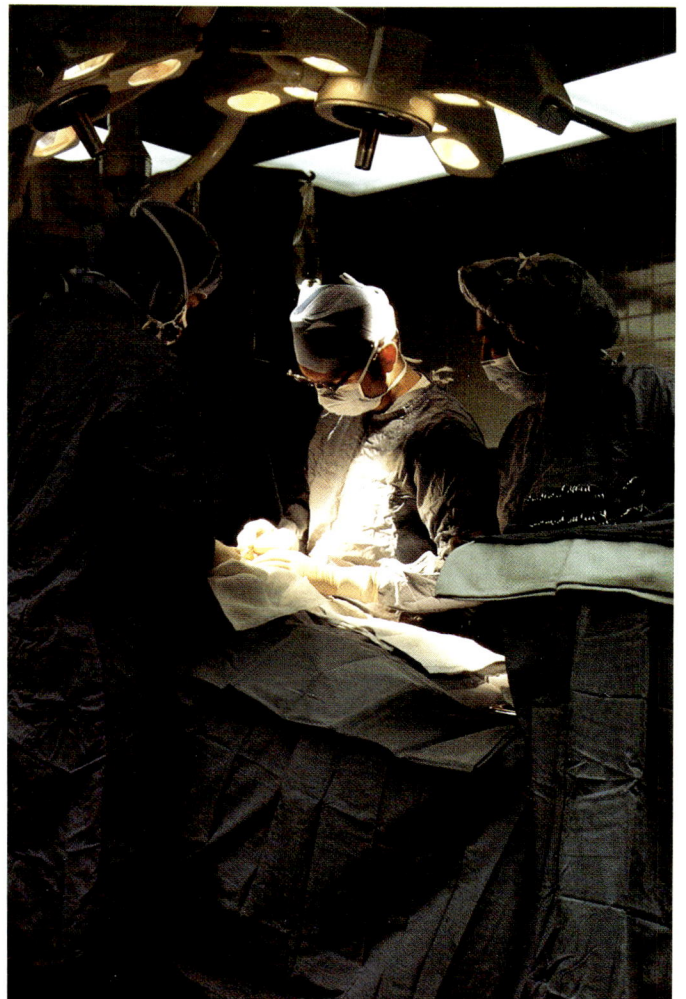

St. Barnabus Hospital maintains the only Burn Unit of its kind within the state of New Jersey. Photo by Sharon Sullivan

St. Mary's, the first public hospital in the state of New Jersey, was established prior to the Civil War in 1863. Photo by Sharon Sullivan

In the 1986 annual report for Overlook Hospital, President and Chief Executive Officer Thomas J. Foley outlined some of the advantages of cooperative efforts. Under a not-for-profit joint holding company known as Atlantic Health Systems, Inc., Overlook and Morristown Memorial have benefited. "Both hospitals were to retain their own boards of trustees and medical staffs," Foley said. But the new entity created several opportunities. Commenting on some of the advantages, Foley points out that there is "strength in size, politically and economically. The system offers more clout in negotiations with state and federal bureaucrats. By the same token, the purchasing power of a 1,300-bed system is far superior to two hospitals half that size." Foley added, "There is also the advantage of maintaining market share, at a time when HMOs, PPOs, and other managed-care programs would be threatening to weaken and divide the two hospitals through competition."

Atlantic Health Systems also prevents the duplication of major capital equipment. "While both hospitals may not have the latest piece of equipment," Foley stressed, "at least one member of the system will." Efforts of this kind and others are being undertaken throughout the state as hospitals search for innovative ways to retain low cost and quality health care while working against adverse economic and regulatory forces within the industry.

NORTHERN HOSPITALS LEAD THE WAY

Over the years, New Jersey's network of fine hospitals has accumulated praise for a wide array of innovations, from financial management and patient programs to sophisticated research endeavors. They have been in the vanguard offering solutions to each new development in the ever-changing health care market. Hospitals, especially in the northern counties, have had longstanding reputations for providing quality health care.

Dating back to before the Civil War, St. Mary's in Hoboken was the state's first public hospital. Begun in January 1863 by four Franciscan Sisters of the Poor and one postulant, the hospital's good sisters ministered to 13 patients and took care of 20 children during its first year of existence. Within months of the establishment of St. Mary's, St. Francis Hospital was founded in Jersey City. In a short time, New Jersey's northern towns were well on their way to charting the course others would follow throughout the state. Even today, the many fine hospitals in northern New Jersey lead the way with innovative programs that have set the standard for institutions throughout New Jersey, and in some cases across the country.

Clearly, University of Medicine and Dentistry of New Jersey is the "jewel in the crown" of New Jersey's health care system. Through its excellent reputation in teaching, technology, and research, it has served to bolster quality health care, which has affected the whole state. The hospital industry is inherently linked with the university medical teaching system through UMDNJ. Its four core teaching centers are geographically distributed throughout the state: UMDNJ-University Hospital in Newark; Robert Wood Johnson University Hospital in New Brunswick; Kennedy Memorial Hospitals-University Medical Center in Stratford; and Cooper Hospital/University Medical Center in Camden.

More than 60 hospitals in the state have one type of affiliation or another with UMDNJ. Through many of these relationships, the university has developed specialized programs at Hackensack Medical Center and other facilities in the north. For example: lithotripsy at St.

Overlook Hospital in Summit is considered to be a pioneer in the field of coronary care, opening one of the first coronary units in the country in 1965. Photo by Sharon Sullivan

HACKENSACK MEDICAL CENTER

The original 12-bed hospital, purchased for $4,000 in 1888, is shown with a horse-drawn ambulance in the foreground. Photo circa 1890

In March 1888 a great blizzard buried most of the eastern United States, along with all of the New York metropolitan area, including a small town on the west bank of the Hackensack River in Bergen County, New Jersey.

Unable to find secure footing on the ice, a brakeman on the Susquehanna Railroad fell on the Hackensack railroad platform and severely injured his head. The nearest hospitals were in other counties, and, according to newspaper accounts at the time, the man's death was blamed in part on the delay in reaching Paterson.

With a young widow and small child left behind, local sympathies ran high for the creation of a hospital in Hackensack. On June 13 a 10-room residence, purchased by an association of 24 civic leaders for $4,000, officially became known as Hackensack Hospital.

What began as a 12-bed facility in 1888 after The Great Blizzard is today Hackensack Medical Center, a 529-bed, regional care, teaching hospital that offers a number of specialized services unique in Bergen County as it celebrated its 100th anniversary in 1988.

The medical center now operates the fourth-largest open-heart surgery program in the state, with a state-of-the-art $2.4-million cardiac intensive care and

step-down unit for patients just out of open-heart surgery, and a modernized cardiac catheterization laboratory that features biplane diagnostic equipment. Nationally recognized cardiac surgeon John E. Hutchinson III, M.D., who performed heart surgery on tennis star Arthur Ashe, thereby helping him return to a full and productive life, recently completed the medical center's 1,900th open-heart surgery.

Construction was begun in 1922 for an enlarged Hackensack Hospital, which opened September 15, 1923.

Hackensack Medical Center has the largest and most comprehensive program in New Jersey for children with cancer. Michael Harris, M.D., and Michael Weiner, M.D., transferred their program from Mount Sinai Medical Center in New York to Hackensack Medical Center in 1987, bringing more than 200 children with them to the program, called The Tomorrows Children's Institute for Cancer and Blood Disorders. A 4,800-square-foot outpatient clinic and 24-bed inpatient unit both feature ultramodern design and the latest equipment.

The medical center also has a nationally recognized program led by Marvin Gottlieb, M.D., called the Institute for Child Development (ICD), an evaluation and treatment program for children with developmental and behavioral problems that attracts children and young adults nationwide and worldwide. The ICD has the only children's hearing program in a Bergen County hospital that is fully accredited by the American Speech-Language-Hearing Association. In addition, the medical center has merged with South Bergen Hospital and uses part of the Hasbrouck Heights facility for its programs.

In 1974 the medical center opened the state's first burn treatment unit, which today is the only burn service in

the Bergen-Passaic area. The medical center also serves as the regional paramedic dispatch center for all the hospitals in northern New Jersey. In addition, the medical center's CPR Training Center is busier than that of any other hospital in the county.

The medical center has the only unit in the county dedicated solely to the treatment of diabetes—the first of its type in the state when it opened in 1974.

Other services include a newborn intensive care unit, limb replantation service, a genetics counseling service, a regional dialysis service, and a Comprehensive Cancer Program affiliated with Memorial Sloan-Kettering in New York City that includes a pain service and the first Medicare-accredited hospice program in Bergen County. Hackensack Medical Center also maintains a virology laboratory that is accredited by the Commission of Laboratory Accreditation of the College of American Pathologists.

A New Jersey State Police helicopter lands at Hackensack Medical Center, whose Emergency/Trauma Department treats approximately 100 people per day.

Left to right: John P. Ferguson, president of Hackensack Medical Center; Stanley Engleman, chairman of the medical center's board of governors; J. Fletcher Creamer; and John E. Hutchinson III, M.D., chief of cardiac surgery, at the medical center's new $2.4-million cardiac intensive care and step-down unit. Creamer, a developer, donated $500,000 toward construction of the new unit.

The medical center is Bergen County's fourth-largest employer, with 2,700 workers. As a major affiliate of the University of Medicine and Dentistry of New Jersey since 1974, the medical center trains some 250 medical students and 80 resident physicians annually. There are more than 500 physicians on staff, representing many different medical specialties. The facility also employs more than 800 nurses.

In recent years the medical center has placed a strong emphasis on outpatient programs. One-third of all surgical procedures performed at the medical center are done without patients having to spend a night in the hospital.

In addition, Hackensack Medical Center reaches out to the community through such programs as nutrition counseling, sports medicine, home health care, phobia and panic disorder clinics, and a Parent-Child Health Program. Home Health Services provides home-care services to acute, chronic, and terminally ill patients, and conducts

health screenings and educational services at community sites. The Health Awareness Regional Program, a multifaceted health promotion and illness prevention program for adults, regularly presents lectures, programs, and screenings as part of an ongoing health education series to help members of the community stay medically informed.

"From its founding as Bergen County's first hospital in 1888, Hackensack Medical Center has grown to become a truly regional facility, providing the highest-quality medical care to New Jersey and the metropolitan area," says John Ferguson, president of the medical center. "We are proud of all our physicians, nurses, employees, volunteers, auxiliary, and board members who have touched the lives of so many for more than 100 years."

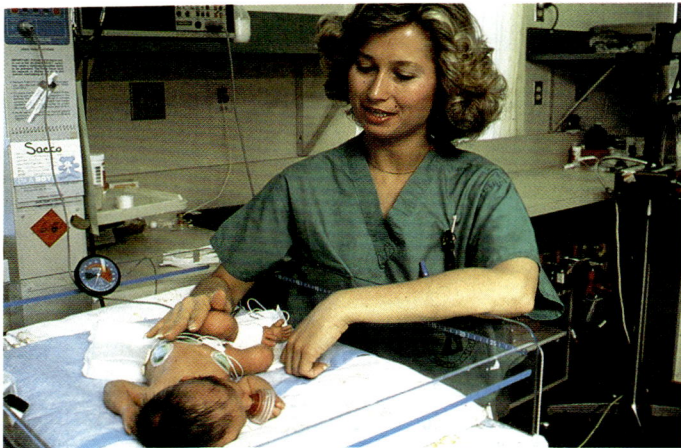

Newborns being monitored receive extra special attention. Courtesy, Overlook Hospital, Department of Public Relations

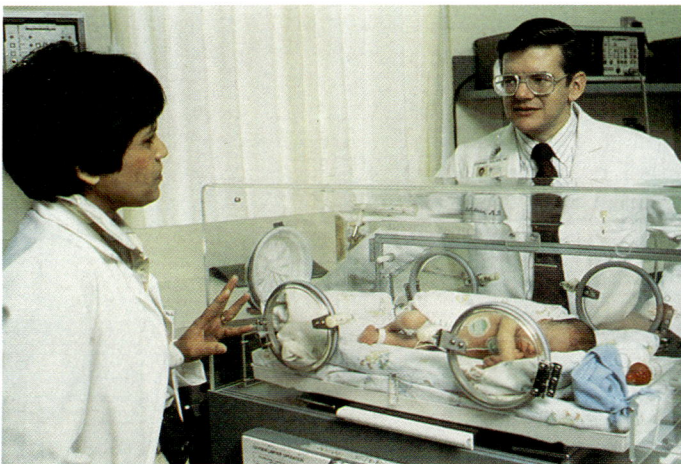

Babies born prematurely generally require an incubator and TLC. Courtesy, Overlook Hospital, Department of Public Relations

Barnabus Medical Center in Livingston; a variety of pediatric special-ties at United Hospitals Medical Center in Newark; and a National Spi-nal Cord Injury Center operated jointly by University Hospital and Kessler Institute for Rehabilitation.

Founded in 1970, UMDNJ has come a long way in a short time. Its 3,000 medical, dental, and allied health students, and the more than 1,000 graduate physicians and postdoctoral fellows, make UMDNJ the largest health science university in the nation. Its prodigious research endeavors have placed both UMDNJ-New Jersey Medical School and UMDNJ-Robert Wood Johnson Medical School among the top four medical schools in the nation in increased percentage of Federal sup-port for research. One of its many ambitious undertakings in biomedical research takes place on UMDNJ's Newark campus. The Center for Molecular Medicine and Immunology (CMMI), a private institute for cancer research, works cooperatively with University Hos-pital's Cancer Center and researchers at the medical school. Estab-lished on the campus in 1983, CMMI is the first center completely devoted to obtaining practical results in human applications in as short

a time as possible in immunology and nuclear medicine research. CMMI receives patient referrals for new cancer detection research and therapy from all over the world.

The Stone Center of New Jersey represents a progressive health care service offered by UMDNJ's University Hospital and St. Barnabus Medical Center. The center offers the most advanced lithotripsy equipment in the country. It is the first of such services to be offered to the physician and the community. Extracorporeal Shock Wave Lithotripsy, performed with improved bath-free equipment, uses water-borne shock waves to crush kidney stones. The tiny particles that remain can be safely excreted. The center is the first facility in the United States to offer the new bath-free lithotripter.

University Hospital is northern New Jersey's premier tertiary care hospital, and houses several services that have no peer. Its New Jersey Cancer Center is one of only six facilities in the United States, and the only one in the New York/New Jersey area, to offer intraoperative radi-ation therapy. In otolaryngology, it has one of the country's first cochlear implant programs. The hospital's Sammy Davis, Jr., National

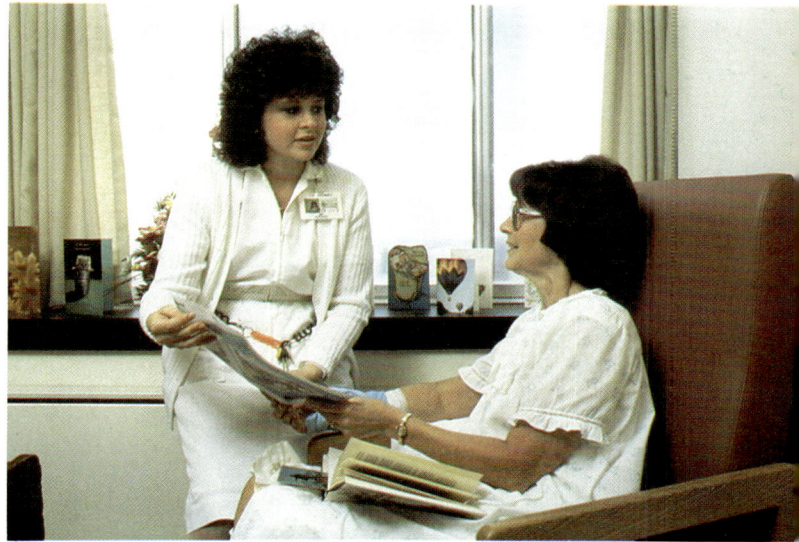

This patient receives assistance with her menu-planner from a nurse. Courtesy, Overlook Hospital, Department of Public Relations

Overlook Hospital was the first in New Jersey to provide a specialized training program for nurses in the field of cardiac care. Courtesy, Overlook Hospital, Department of Public Relations

Liver Institute will serve as a national resource for liver disease research and treatment. It is the country's first medical facility devoted solely to patient care, education, and research for liver disease. The hospital also houses the New Jersey State Trauma Center, which provides state-designated Level-I trauma care for all of northern New Jersey. High-risk pregnancies and newborn follow-up are provided through the university's statewide Perinatal Services and Research Center. The center has been given the highest designation: super-level III. The state's only neurosurgical intensive care unit is also located at the Newark facility. The hospital is also known for its advanced program of orthopedic surgery and sports medicine, and its unique pain management program.

All of these specialized services available at UMDNJ's University Hospital create the highest level of health care found in the state, and more and more patients from out of the county and out of the state are being referred to the northern New Jersey facility for the incomparable care it can provide. The facility has helped to make northern New Jersey a mecca for sophisticated health care.

Northern New Jersey is blessed with possessing many first-rate hospitals, most of which have well defined areas of specialization. What follows are a few noteworthy examples:

—St. Barnabus, the oldest and largest acute care hospital in the state, provides several outstanding services in addition to its Stone Center. Its twelve-bed Burn Unit is the only one of its kind in the state. Its radiotherapy program is the largest in northern New Jersey. The hospital is one of only five regional oncological centers in the Northeast.

—Overlook Hospital in Summit has had a long history of involvement in cardiac care—in fact, it has been a participant from the beginning as a pioneer in the field. Overlook's coronary care unit was the first in New Jersey, opening its doors on April 16, 1965. At that time, it was one of only 12 coronary units in the nation. Not only was the unit the first in the state to assemble special equipment and treatments, but it was the first to offer specialized training in cardiac care for nurses. In the interim, the hospital has built on its initial reputation. The cardiology section of the Department of Medicine is one of its

St. Clare's Riverside Hospital is well recognized for its "Living With Cancer" programs. Photo by Sharon Sullivan

strongest and most active groups. Heavily involved in research relating to coronary heart disease since 1975, Overlook was the only community teaching hospital selected to participate as a clinical center in four multi-center international clinical trials. It has also participated in clinical research studies in cardiology sponsored by Ciba-Geigy, Sandoz, and Hoechst Pharmaceutical Industries. Overlook is also a pioneer in neurosurgery, and for more than 25 years has been a recognized leader in diagnosis and neurological disorders. The hospital's technological commitment to neurosurgery began in 1959 when it acquired the nation's first electroencephalograph. The hospital also introduced the state's first computerized axial tomography (C.A.T.) scanner, and the state's first intensive care unit.

—Fair Oaks Hospital in Summit has the distinction of getting the public's attention in a "big way." In 1983, the hospital's National Cocaine Hotline gained national attention when its endeavors were heralded in *Time Magazine,* most of the nation's major newspapers, and the broadcast media across the United States. Its "800-Cocaine" toll-free telephone number connects a cocaine user, family, or friends from anywhere in the nation to a staff of cocaine rehabilitation experts at the hospital. As the first program of its kind, it was largely responsible for revealing the depth and breadth of the nation's cocaine dependency problem. Fair Oaks has gained an international reputation both as a psychiatric hospital and as a drug-abuse treatment and research facility.

Quality health care utilizing the most advanced, efficient methods is readily available in New Jersey's northern counties. Courtesy, Overlook Hospital, Department of Public Relations

grams. Started in February 1980, the cancer patients and their families meet weekly in groups to discuss their problems, concerns, and diagnosis. The oncology nursing staff and a consultant and trainer from the consultation and education department of the Community Mental Health Center assisted during meetings. Eventually other groups evolved as the need arose: families meet without the presence of the patients; children and teens meet in their respective groups to discuss problems unique to their age groups; and bereavement groups designed to address the needs of each age classification meet to help members work through the grief of losing a family member to cancer.

St Clares children's groups are unique. Begun in 1982, they were the first such groups in the country. In an article in the *New York Times,* Barbara Blumberg, public health educator of the National Cancer Institute, commented on the need for such a service: "Suddenly everyone seems to realize how incongruous it was that kids were practically left to fend for themselves." The program at St Clares • Riverside constitutes a landmark step toward confronting the problem. Inquiries have come in from across the country. The hospital's bereavement groups are also the first to offer support after the patient's death. The programs run weekly, and they are available to troubled families throughout northern New Jersey.

—Pascack Valley Hospital in Westwood is one of the first hospitals in the state to have a division of sports medicine within its department of medicine. The department is headed by Dr. Allan Levy, the medical director for the Giants, New Jersey's football team. He has also had affiliations with the New Jersey Nets basketball team and the Devils, the state's hockey team. Levy treats only patients with sports-related injuries, and the hosptal's program takes an aggressive fitness route before resorting to surgical alternatives.

Pascack Valley is also one of the first to offer "birthing suites." Constructed in 1985, the birthing suites were established to fulfill a need expressed by patients seeking an alternative to the sterility of traditional childbirth methods. The two suites each include a birth room and an anteroom with kichenette, couch, and bathroom where

—Founded in 1888, Hackensack Medical Center in Bergen County has the largest pediatric cancer service in the state. Two nationally recognized leaders in the field of childhood cancer direct the new program: Dr. Michael Harris, chief of pediatric hematology-oncology at Mount Sinai Medical Center, and Dr. Michael Weiner, associate chief. According to John P. Ferguson, president and CEO of the medical center, "with the addition of pediatric oncology, the hospital has the most comprehensive program for the diagnosis and treatment of children and adolescents with cancer and blood disorders in all of New Jersey." Pediatric oncology completes the range of services offered by the hospital's comprehensive Cancer Program.

—St Clares • Riverside Medical Center, a 401-bed suburban institution with divisions in Denville and Boonton, has also gained a great deal of press attention for its innovative "Living With Cancer" pro-

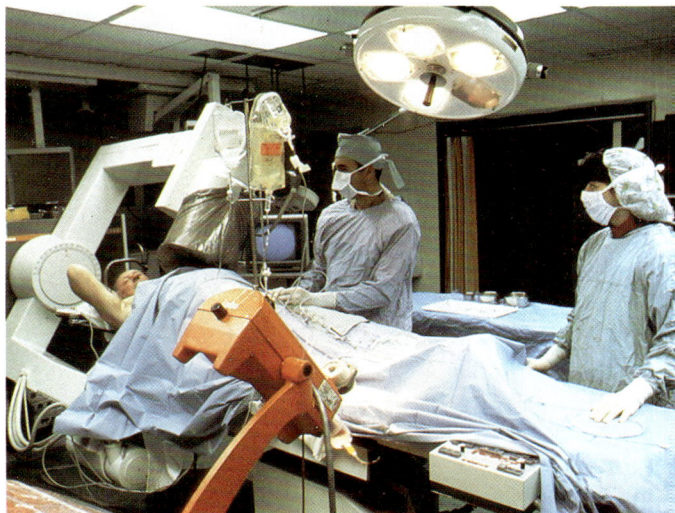

Approximately 1,100 open-heart operations are performed annually at Beth Israel Hospital in Newark.
Photo by Sharon Sullivan

family members can stay. The hospital also offers the option for a newborn to be delivered by a midwife. Pascack Valley is one of the few hospitals in the state that has midwives who are credentialed staff members with admitting privileges. The midwives are independent practitioners working in homes, birth centers in the area, and the hospital, where medical backup is available.

— Beth Israel Hospital in Newark is the only hospital in the state that performs heart transplants. Established in 1985, it is part of an extensive cardiac diagnosis and treatment program. The program is one of the busiest in the state, with approximately 1,100 adult open-heart operations performed a year.

EMBRACING THE FUTURE   Northern New Jersey's hospitals have coped admirably with the myriad changes within the health care industry. However, the region's health care institutions are also preparing for the future. In many cases that means redirecting focus on outpatient care programs. What follows is only a brief overview of a few notable accomplishments and some future plans:

— A continuing care retirement community will be constructed on grounds adjacent to the Denville facility of St Clares • Riverside. It will include a 60-bed skilled nursing facility and an independent living complex with 260 units. There will be 40 two-bedroom country homes plus 220 one- and two-bedroom units. The hospital is a result of a 1985 merger between St Clares, founded in 1953, and Riverside, founded in 1955.

— In 1986, Wayne General Hospital dedicated a new five-story wing. The new area houses a 34-bed medical/surgery unit, a new intensive care/critical care unit, an all-new and larger lab, EEG/EKG departments, respiratory therapy facilities, and a pharmacy. Wayne General has a long heritage of service to the community since 1871.

— Morristown Memorial's $19-million "New Era of Excellence" Campaign supports the expansion of its outpatient and regional referral services and the restructuring of acute care facilities, which are part of a $37-million development program. Morristown Memorial's Center for Geriatric Care has begun a respite program that provides skilled nursing for the elderly for one day to two weeks.

— Kessler Institute for Rehabilitation has responded to the trend toward outpatient services. Kessler will add satellite facilities beyond its main facilty in West Orange. Kessler's other northern locales are in East Orange, Saddlebrook, and Union County.

— Founded in 1873, the Hospital Center at Orange offers a program of home health care for community residents. HCO Plus ministers to residents whether they have been hospitalized or not. A num-

Recent trends lean toward a specialization in hospital facilities. One hospital in northern New Jersey offers sibling education for expectant mothers with older children. Photo by David Greenfield

ber of home care services are provided by home health aides, registered and licensed practical nurses, social workers, nutritionists, and occupation and speech therapists.

—In 1985, the General Hospital Center at Passaic completed a $70-million reconstruction program and a seven-story facility, allowing expansion of the hospital's cardiac services. A $10-million radiology department opened in 1986.

These and many more hospitals in New Jersey's northern counties have or plan acquisitions, renovations, or expansions, always keeping in step with the best and most efficient ways to deliver health care services.

The number and quality of the institutions in northern New Jersey makes access to quality care relatively easy. Bergen County alone has nine general private hospitals and one county hospital serving its residents. Passaic County, which is moderate in size, has seven major hospitals within its boundaries. Each of these facilities provides excellent medical care and skills and the most advanced equipment available. Residents of the northern counties are also fortunate to have ready access to the medical, surgical, and health research resources of nearby New York City.

Meadowlands Hospital provides excellent care for local residents. Photo by Sharon Sullivan

The proliferation of responsive and vibrant health care institutions makes the northern portion of the state unsurpassed in excellence. Today, as in the past, northern New Jersey hospitals deal successfully and imaginatively with the rapid-fire changes within the industry.

Northern New Jersey airports are a pleasant alternative to the congestion of nearby major airports. Photo by Frank Oberle

Northern New Jersey possesses one of the most highly developed and sophisticated systems of transportation in the nation. Occupying one of the the country's prime travel corridors, networks of highway, railroad, aviation, marine, and pipeline facilities literally operate side-by-side, forming a unique transportation matrix, especially in the Newark International Airport area. An unusually high degree of mobility is supported by the commitment of New Jersey's northern counties to sustaining efficient systems. Roadways and rail lines allow people to work, shop, socialize, and conduct business within the region, and its fine airports and ports make other parts of the country and the world readily accessible. Northern New Jersey's preponderance of national

CHAPTER
ELEVEN

and international corporations prove daily that doing business here is easier than almost anywhere else. Northern New Jersey's vast transportation network has not only spurred, supported, and sustained the region's phenomenal growth and development but has made it a major international transportation nexus as well.

HIGHWAYS AND BYWAYS In 1775, the news of the Revolution's start was carried by a courier on horseback, from New Haven,

The Transportation Nexus Of The Northeast

Route 3 is an eight-mile stretch of turnpike connecting Clifton and North Bergen. Photo by Bob Krist

Connecticut, to Georgetown, South Carolina. In a historical account, Robert Greenhalgh Albion reports, "Good statistical evidence indicates that well-traveled New Jersey had the best road system in the colonies at the time of the Revolution."

The state's contributions to road building since the nation's fight for freedom have not abated. In his *Gems of New Jersey,* Gordon Bishop lists the state's accomplishments:

—The nation's first paved road built by the Dutch in 1664, in Warren and Sussex counties.

—The first public road act in 1673.

—The first road-sheet asphalt pavement, in Newark in 1870.

—The first traffic circle, in Camden in 1925.

—The first cloverleaf.

—The first concrete divider in the middle of highways, dubbed the "Jersey Barrier."

—The first state to grant monetary aid to build public roads, in 1891.

Northern New Jersey highways still maintain a reputation of excellence, moving people and cargo from point to point with relative ease. Automobiles, trucks, and buses make their way within the state over thoroughfares that thread gracefully through urban areas and verdant suburban landscapes alike. Routes 202 and 206 meander through the state, offering up examples of its beauty. The Palisades Interstate Parkway is perched atop the dramatic Hudson River Palisades. Old state roads like routes 10, 14, 45, and 57 still service the public daily. Routes 4 and 17 occupy strips along which reside the most spectacular array of shopping centers in the United States. Multitudes of faithful shoppers make pilgrimages to this commercial mecca throughout the year. Likewise, routes 22, 46, and 1-9 vie for first place as the most congested consumer byways in the north.

Any or all of these roads are pathways by which the avid traveler can fully experience northern New Jersey's infinite variety: cities and suburbs, forests and farmlands, rugged mountains and jagged cliffs, rivers and lakes, beautiful meadowlands, and the world's most famous

The New Jersey Turnpike officially opened as the state's first modern toll road in 1952. Photo by Carol Kitman

port. In addition, two toll roads have received national notice. The New Jersey Turnpike is the busiest and safest toll road in the country. It is also the widest road in America, with some stretches accommodating 12 lanes. The Garden State Parkway, on the other hand, is as beautiful as it is utilitarian. This lengthy and scenic highway links the northern counties to the shore areas and Atlantic City. The interstate routes act as a link to other parts of the country. For example, Route 80 traverses the nation and ends in California.

MAINTAINING A SOUND INFRASTRUCTURE In 1986, the New Jersey Department of Transportation (NJDOT), celebrated the twentieth anniversary of the implementation of the Transportation Act of 1966, which laid the groundwork for the department's creation. During the year, projects of great import to the health of the state's transportation network were begun, and many completed. More roads and bridges were repaired and rebuilt and more missing highway links were completed than ever before. According to the 1986 annual report of the New Jersey Transportation Trust Fund Authority, "this major overhaul of New Jersey's $42-billion transportation infrastructure has generated thousands of jobs, while stimulating economic growth."

During 1986, NJDOT orchestrated a series of projects that affected several northern New Jersey highway networks. Major construction led to the completion of the missing I-78 link in Union County. The new $111-million, six-lane highway connects the 18.8-mile section from Springfield east to New York (Holland Tunnel) and the 38.3-mile section from Berkeley Heights west to Still Valley, near the Pennsylvania border where the last segment of the interstate in New Jersey is under construction. According to its report, NJDOT claims that these stretches of highway include "more environmental safeguards than any other road construction project ever undertaken in New Jersey" —a result of the long deliberation the project was given. Although it took only four years to build the new link, it had been the subject of environmental studies, public hearings, and legal action since 1972. Construction was also begun on the final 21.1 miles of I-287 from Montville to the New York State Thruway. The $400-million extension is expected to be completed in 1992.

Traffic circles, a New Jersey invention, have been meeting their demise as NJDOT embarked on eliminating the most dangerous ones.

Interstate 78 through Hunterdon proves to be, time and again, a scenic and relaxing drive. Photo by Michael Spozarsky

The Art Deco patinated Hoboken Terminal is a noted landmark among New Jersey residents. Photo by Carol Kitman

Many miles of northern New Jersey's train tracks are beautifully landscaped and well maintained. Photo by Rich Zila

The platform of the Hoboken Train Terminal is abstracted by the light of day. Photo by Michael Spozarsky

Once considered an engineering model for safety, the circle has since outlived its usefulness. As traffic increased, so did the hazards of negotiating the state's many circles. In its campaign, the department eliminated three circles in the north: two in Morris County and one in Bergen. In Morris County, the Route 23 Riverdale Circle was torn up, and work commenced on eliminating the Route 23 Jackson Circle. In Bergen County the $1.6-million project to eliminate the Lodi Circle will be completed this year.

The department has also embarked on work on the three bridges needed for the new 6.1-mile Route 24 Freeway extension in Morris County. In Bergen and Hudson counties, routes 4, 7, and 17 are undergoing major improvements. NJDOT also continued its extensive $128-million program to rehabilitate more than 100 bridges on I-80 between the George Washington Bridge in Bergen County and the Delaware

Water Gap. Rehabilitation has begun on I-80 bridges in Totowa, Paterson, West Paterson, and Elmwood Park, in Passaic and Bergen counties. Interchange revisions and bridge rehabilitation has also been undertaken on routes 1-9 in Essex and Hudson counties.

In its ongoing commitment to keep in step with the state's rapid economic growth, NJDOT has sought to define and meet the state's future transportation infrastructure needs. Improvements continue throughout the state, keeping New Jersey in the forefront of a major highway revitalization.

RAILROADS AND BUSES   New Jersey's litany of firsts did not cease with highway advancements. Colonel John Stevens of Hoboken pioneered the development of rail travel in New Jersey as early as 1812, when he petitioned the commissioners of the Erie Canal requesting $3,000 to conduct an experiment to prove that rail shipment was more efficient than canal transportation. It was only fitting for Stevens to become involved with the project since he had launched the world's first steam ferry on the Hudson River. His belief was that freight could be shipped across land more cheaply on a steam-powered railroad. Stevens managed to obtain the first railroad charter in the United States on February 6, 1815. A decade later, he ran the nation's first steam-operated locomotive, a "steam wagon," on a circular track in Hoboken. Robert L. Stevens, the colonel's son, continued the family tradition when he secured the state's first commercial railroad charter in 1830. He is also responsible for designing the first "T" shaped rail, the "hook-headed" spike used to fasten rails to ties, and the "iron tongue"

Above: Cables which support the George Washington Bridge lend a graphic element to this image of commuters returning to Fort Lee. Photo by Michael Spozarsky

used to join rails.

Nearly 160 years later, New Jersey's railroads continue to operate efficiently every day. New Jersey Transit, New Jersey's public transportation corporation, oversees rail and bus passenger service throughout the state. It is one of the largest public transportation operators in the nation, providing more than 600,000 passenger trips daily. Its rail links dozens of communities with Newark, Hoboken, and New York City, and its extensive network of bus lines operates throughout New Jersey and provides service to New York City, Philadelphia, and Wilmington, Delaware. Its rail lines connect more than 140 cities and towns, and 189 bus routes run through urban centers and suburban communities.

Since its creation in 1979, New Jersey Transit has worked to keep pace with the constant demands of the state's rapid and unabated growth, which occurs especially in the northern counties. The economic boom and its concomitant growth in the housing market has also made trans-Hudson public transportation links essential. The Port Authority Trans Hudson, or P.A.T.H. (operated by the Port of New York and New Jersey Authority), serves thousands of commuters daily. The P.A.T.H., along with Amtrak and New Jersey Transit, put Manhattan only a short train hop away, making employment in New York City a relatively easy matter for New Jerseyans.

With an eye fixed firmly on the future, New Jersey Transit has plans for route expansions and additions to its existing services and equipment. In fiscal year 1986 alone, New Jersey Transit added in excess of 9,000 seats to its rail fleet through rehabilitation of old cars

Above: Pennsylvania Station in Newark is a constant nexus of activity. Photo by Sharon Sullivan

Left: Signs for PATH (Port Authority Trans Hudson), such as this one located in downtown Hoboken, are a familiar sight to New Jersey residents. Photo by Carol Kitman

The Newark International Airport occupies approximately 68 acres of land which had originally been bogs and swampland. Satellite extensions of main terminals, connected by covered walkways, are part of the more recent development program. Photo by Michael Spozarsky

and the acquisition of new ones. Its bus system expanded as well, with 3,000 added seats for peak hours. In addition to 250 new commuter buses, New Jersey Transit plans to refurbish 12 buses and acquire 110 new articulated buses—these "dependable bendables" are extra-long buses that have 40 percent more seats than conventional buses. The corporation's new services have been popular. In particular, its package train-and-bus service to and from the New Jersey shore gives northern New Jerseyans a trouble-free means to get to their favorite beaches. Its future considerations also include the increased demands for public transit to New York City, the Hudson River waterfront, and other growth areas in the state.

AVIATION New Jersey has been a major player in air travel since its inception. In fact, New Jersey's long aviation history parallels the growth of aviation in general in this country. One of northern New Jersey's most interesting museums attests to that fact. At the Aviation Hall of Fame Museum at Teterboro Airport, New Jerseyans and other Americans as well can trace the state's and the nation's progress from the first attempts at flight to space exploration. The list of New Jersey people and places involved in advancing aerial exploration is impressive. According to Bishop's *Gems of New Jersey,* they include:

—The first aerial landing in the Western Hemisphere in 1793, when Jean-Pierre Blanchard, a French adventurer-balloonist, arrived in Woodbury.

—The first airmail letter delivered in America, delivered to New Jersey by Blanchard. It was a missive from George Washington to be presented to the first person Blanchard met.

—The first manned parachute descent in the United States, by Louis Charles Guille in 1819 in Jersey City.

—The first native American to fly, in a balloon in 1830, by

Newark International Airport officially opened in 1928. It is the oldest commercial airport in the New York Metropolitan area. Today the airport services nearly 30 million customers yearly. Photo by Michael Spozarsky

Newark Airport maintains approximately 1,100 flight operations daily. Photo by Bob Krist

Charles Ferson Durant, a resident of Jersey City.

—The 1885 flight of Lucretia Bradley, the first American woman to fly.

—The first pilot to reach an altitude of more than one mile—Richard Brookins' 6,176-foot-high flight in 1910.

—The first municipal airport in the world, opened in 1919 in Atlantic City at Bader Field.

—America's first terminal for international passenger service by dirigible, established in Lakehurst. By 1919, Lakehurst Naval Air Station became known as the "Lighter-Than-Air Capitol of the World." It served as the home base for every Navy dirigible and the German trans-Atlantic airships *Graf Zeppelin* and *Hindenberg*. The *Hindenberg* met its famous fiery end in 1937 at Lakehurst. The disaster also ended the era of the great Zeppelins.

—The world's largest, single-wooden-arch hangars, at Lakehurst.

—The first supersonic flight, accomplished by Captain Charles (Chuck) Yeager in a rocket engine developed and built by Reaction

Motors in Rockaway in Morris County.

As an example of some of the accomplishments New Jersey and its citizens have been a part of, this list represents only a few of the highlights.

New Jersey's impressive strides in aircraft manufacturing also obtained world attention. Notably, Wright Aeronautical Corp. and Wright-Bellanca Co. of Paterson, and Fokker Aircraft Corp. of Teterboro, lead the field with their designs. Most recorded historic flights during the second and third decades of the twentieth century were accomplished with planes designed in New Jersey. Even Charles Lindbergh's *Spirit of St. Louis* had a Wright Whirlwind engine propelling it.

New Jersey's enthusiastic participation in the growth and development of air travel has been unabated from the outset. It is only fitting, then, that New Jerseyans should lead the way into outer space. Walter Shirra of Bergen County was not only one of the first "team" of astronauts, but he is the only American to have done it in three different space crafts: Mercury, Gemini, and Apollo. Another northern New

THE PORT AUTHORITY OF NEW YORK AND NEW JERSEY

On April 30, 1921, as the first of its kind in the Western Hemisphere, the New York-New Jersey Port Authority came into being. The new agency's area of jurisdiction was called the Port District, a 17-county bistate region within a 25-mile radius of the Statue of Liberty.

The mandate of the agency was to promote and protect the commerce of the bistate port and to undertake port and regional improvements not likely to be invested in by private enterprise nor to be attempted by either state alone—a modern wharfage for the harbor the two states share, tunnel and bridge connections between the states, and, in general, trade and transportation projects to promote the region.

In keeping with its mandate, the New York-New Jersey metropolitan region completed its 11th consecutive year of economic growth in 1987, matching or outpacing by most measures an expanding national economy.

The Port Authority, with the encour-

The Port Authority of New York and New Jersey's area of jurisdiction covers a 17-county bistate region within a 25-mile radius of the Statue of Liberty. Today the Port Authority plays an active role in stimulating regional renewal and redevelopment.

agement of the two states, has broadened its mission to include playing an active role in stimulating regional renewal and redevelopment. Since 1978 the agency has committed hundreds of millions of dollars to projects that would help rekindle commercial development, renew aging infrastructure, and attract new industry in the region.

The Port Authority faces new challenges that can be met only by new large-scale commitments. The region's mounting prosperity and its shifting commutation patterns have stretched this agency's basic transportation and trade facilities to their limits. At the same time, other regions are making aggressive efforts to capture this area's

preeminent share of the nation's commercial and transportation activity.

As the Port Authority sustains its singular commitment of agency resources to meet today's and tomorrow's regional needs, it must also continue to fulfill its commitment to regional economic development. At the same time, the Port Authority must focus new energies and resources on expanding and upgrading its essential trade and transportation facilities to prepare them for the demand to come.

To sustain and expand the benefits of regional growth, The Port Authority of New York and New Jersey has enacted with both states a $5.8-billion capital investment program covering both the agency's outstanding commitments and programs, and the major new capital and service improvements to its core trade and transportation facilities over the next five years.

Capital plan project and program categories include renovations; modernizations; expansions and new development in the bistate agency's airports, tunnels, bridges, terminals, and industrial parks; as well as promotion of regional economic development and port commerce. The complex and far-reaching projects the Port Authority is undertaking will extend through this decade and into the next.

Unlike some other government entities, the Port Authority's system of facilities must be self-supporting. While commercial tenants share in the costs of some facility improvements under lease arrangements, implementation of the program depends on availability of adequate revenues from tolls, fares, other user charges, and rents to support both operating costs and debt service on Port Authority bonds—those currently outstanding and those issued to support the capital program.

"This is a bistate effort—not simply because we have by statute certain obligations on each side of the Hudson, but because the area we serve is truly one economic region. The facilities we operate and the services we provide help knit the region together," says Port Authority executive director Stephen Berger.

Port Newark/Elizabeth is the leading container port in the United States. Among its diverse cargoes, the volume of Japanese cars which pass through the port exceeds that of any other U.S. port. Photo by Bob Krist

Jerseyan, Buzz Aldrin of Montclair, in Essex County, was selected for NASA's moon mission; he walked on the lunar surface, collected samples, and conducted experiments.

In 1927, the City of Newark assured its place in New Jersey's continuum of aviation firsts when it proceeded with plans to build the first major commercial airport in the greater metropolitan area (both Kennedy and LaGuardia airports had their beginnings during World War II). Sixty-eight acres of swampland were transformed, and by 1930, Newark's airport was heralded as the "busiest airport in the world." Over the years, many notable aviation achievements have taken place at the airport:

—In 1934, Captain Eddie Rickenbacker landed at the airport from Los Angeles, setting a new coast-to-coast passenger flight record: 13 hours and 2 minutes.

—In 1935, Amelia Earhart flew nonstop from Mexico City to Newark in 14 hours and 19 minutes.

—In 1936, Howard Hughes flew from California to Newark in a record 9 hours and 26 minutes.

Newark International Airport serves approximately 30 million customers yearly with 1,100 flight operations per day. Its 2,200 acres make it the state's largest airport. It is operated by the Port Authority under a lease agreement with the City of Newark. Improvements and expansion have made it the second-busiest airfield in the tri-state area,

and it is rapidly overtaking La Guardia. Quoted in *New Jersey: History of Ingenuity and Industry,* Vincent Bonaventura, general manager of New Jersey airports for the Port Authority, said the airport tripled its passenger usage to nearly 29 million between 1981 and 1986, making it "the fastest-growing major airport in the world."

Newark's airport is joined by a series of fine airports that service northern New Jersey. Teterboro Airport's 878 acres accommodate personal and corporate aircraft. Owned by the Port Authority of New York and New Jersey, the property was originally acquired by Walter C. Teter in 1917. Despite its long and illustrious aviation history, including Amelia Earhart's many flights in New Jersey, one incident in 1954 made Teterboro nationally famous. Radio and television personality Arthur Godfrey, who was born and raised in Bergen County, was charged with "buzzing" the tower and found guilty. His flying license was revoked for six months and the incident was frequently referred to by him on his television program. As a result, Godfrey's demise became part of Teterboro's colorful history. Before the incident, Godfrey had achieved notoriety as a fine flyer. In 1947, he flew solo from Teterboro to Point Burrow, Alaska, and back again. He logged 12,000 nautical miles in 62 hours. Today, Teterboro's famous tower houses its Aviation Hall of Fame. The airport is one of six airports in the state with a control tower, and it handles more than a quarter-million takeoffs and landings a year.

Morristown Airport in Morris County operates on 600 acres. Originally built by the military in 1941, today the airport is designated a "reliever" for Newark. It can accommodate aircraft as large as 747s. In addition to Teterboro and Morristown, northern New Jersey is dotted with small airports and jetports that serve business and private communities.

Tugboats are essential equipment providing smooth harbor operation. These boats are currently drydocked in Jersey City. Photo by Carol Kitman

PORT NEWARK/ELIZABETH SEAPORT From the outset, northern New Jersey has been cognizant of its fortunate position near the greatest port area in the country. (The area was first explored in the seventeenth century when Robert Treat sailed through Newark Bay and up the Passaic River to the vicinity of Mulberry Street in Newark.) In 1914, ground was broken near Peddie Creek to create a 20-foot-deep channel, as officials in Newark prepared to create its port. By 1916, the *A.J. West,* a schooner that had carried mahogany from Manila, became the first ship to unload its cargo at the new seaport. In 1948, the Port of New York Authority assumed responsibility for the operation and development of the port. Eventually construction commenced to convert marshlands south of the port in 1958. The new facility, the Elizabeth-Port Authority Marine Terminal, was created by dredging Bound Creek into Elizabeth channel.

By 1973, Port Elizabeth handled 1,125 vessels and more than 7.3 million tons of cargo. The 1,165-acre facility, specializing in handling containerized cargo, came to be known as "America's container capitol." In 1986, the Port Newark/Elizabeth Seaport began work to deepen Newark Bay to 40 feet, which would allow it to handle modern, giant container ships. As one of the world's most efficient ports, it is also the leading container port in the United States.

James J. Kirk, director of Port Authority of New York and New Jersey's port department, said in November 1987 in *New Jersey Business,* "the Port of New York and New Jersey's marine industry contributes $14 billion annually to the region's economy, and generates nearly 200,000 jobs." New Jersey's early decision to go to container terminals is the reason for that success. "More cargo is moved through a single facility in Port Elizabeth every year than through the entire Port of Baltimore," said Anthony J. Tozzoli, president of the New York Shipping Association, in the same article. He asserts that the port handles more

The Port of Newark opened in 1915 and was managed by the city of Newark until 1948. At that time the Port was leased to the New York Port Authority which spent millions of dollars to completely rebuild the facility. Photo by Bob Krist

than one-sixth of the nations' total ocean-borne foreign trade. In the same article, Kirk added:

The Port region is the nation's greatest consumer market and financial center, with over 15 million people living and working here. It is not surprising that its eleven container port facilities offer more cargo handling capacity than all other major Eastern Seaboard ports combined. It presently provides overnight intermodal delivery to 72 million people using a skilled, readily available work force. And it plays host to nearly 100 steamship lines from around the world.

With its phenomenally efficient transportation network of highways, railroads, airports, and seaports to support it, northern New Jersey couldn't fail to fulfill its destiny to become a major force in America's economy. It is amply equipped with the wherewithal to move people and materials within the region, and it truly offers a gateway to the world marketplace.

Trucks and their transport are permitted on the New Jersey Turnpike. Photo by Michael Spozarsky

An unexpected summer shower is easily accommodated! Photo by Nancy Brown

Despite New Jersey's diminutive size—it is the fifth-smallest state in the nation—its leisure-time offerings are boundless in number and variety. The rugged mountainous country in the north provides the perfect topographical contrast to the sandy expanse of prime coastline in the south. These two diverse areas have been magnets that draw travelers and natives alike, making New Jersey a vacationers' paradise. Figures released by the U.S. Travel Data Center ranked New Jersey fifth in the nation in tourism revenues in 1985. By 1987, according to *New Jersey Success* (December 1987) tourism ranked as the state's number-one industry, generating over $12.9 billion in expenditures annually. Clearly, New Jersey is in the forefront of the nation's tourist trade.

CHAPTER TWELVE

THE NATURAL ENVIRONMENT: TOPOGRAPHY The panoply of opposites that fuels New Jersey's tourist trade is most evident in

The Leisure Life: Where Opposites Attract

Jenny Jump State Park is a virtual haven for the day-hiker. Photo by Michael Spozarsky

the north. The region is a four-season vacation retreat: a microcosm of recreational pleasures that can be found otherwise only through extensive travel. What stands out is the sheer beauty that marks the area's landscape. The Appalachian Mountains cross the northwest portion of the state in a southwesterly direction. The highest of these are the Kittatinny Mountains, which extend along the Delaware River from a peak called High Point (the highest point in the state) to the breathtakingly beautiful Delaware Water Gap. The Highlands, which lie east of the Kittatinnies, are composed of rolling green hills accented with crystal-clear lakes. East of the Highlands extends a rolling valley broken by ridges and single mountains, as the land descends gradually to the meadows of the Hackensack Valley and the coast. From Weehawken to a point north of the New York border, the Palisades form a barricade of stony cliffs that rise 200 to 550 feet above the waters of the Hudson River.

Northern New Jersey's urban areas fall within these regions, making pastoral settings and urban sophistication both readily accessible. And even though northern New Jersey has no direct access to the ocean, the native and traveler alike should not feel deprived. The numerous lakes and streams—there are 110 in Sussex County alone—provide an arena for every kind of water sport. Whether your tastes incline to nature walks along the Appalachian Trail or assiduously pursuing history, northern New Jersey can accommodate your particular taste. You are more likely to exhaust your energies before you exhaust the possibilities.

Refreshing entertainment is available at Action Park. Photo by Michael Spozarsky

SUSSEX COUNTY   Perched atop New Jersey and bordered by New York and Pennsylvania, Sussex County has become a key growth location in the northeast and one of the most desirable places in the country to live. Sussex County's growth has been the second-fastest among New Jersey's counties for the past 10 years. However, despite this tremendous growth pattern, Sussex County, the fourth-largest county in

York's Catskills, Sunrise Mountain in Stokes State Forest, Sawmill Lake, and Lake Rutherford, the source of water for the town of Sussex. The obelisk also offers a bird's eye view of Lake Marcia, a lake at the highest elevation in the state.

Montague, the northernmost town in New Jersey before reaching the Pennsylvania border, has a unique history; it has been both the Delaware Indian capital of Minisink and an isolated early Dutch village. By the time the English explorers came to the area, the Dutch spoke only Lenni-Lenape.

Waywayanda State Park is located in the northeast corner of Sussex County, 11 miles northeast of Hamburg. The 9,075-acre mountain woodland tract is under development, but excellent fishing may be enjoyed in the 165-acre lake. Canoeing, paddle boating, and picnicking are also possible.

With the impressive mountains in Sussex County, it's no surprise that this part of the state is a popular ski area. The bulk of skiing is in Vernon Township, considered by many as the ski capital of New Jersey. Vernon Valley/Great Gorge ski area has nearly 60 trails ranging in degrees of difficulty from beginner to expert. Trails drop from as high as 1,000 feet. Hidden Valley ski area has a vertical drop of 620 feet. As

A cool retreat from summertime heat: Tillman Ravine in Stokes State Forest. Photo by David Greenfield

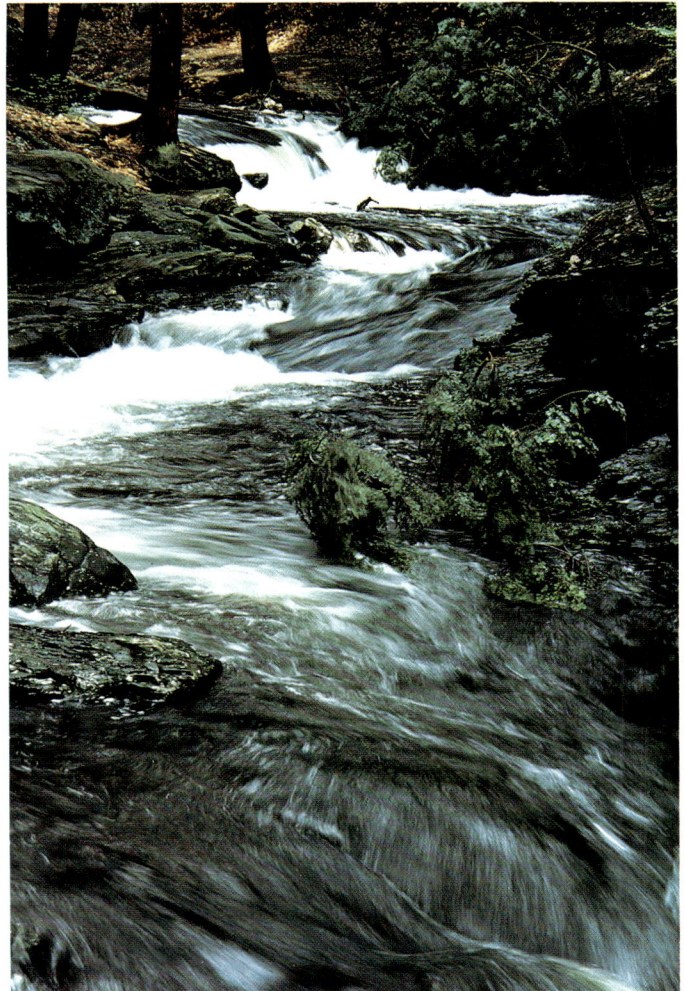

Brightly colored hot air balloons are easily sighted against the green, rolling farmlands of Sussex County. Photo by Bob Krist

the state, will still be one of its least densely populated. With over 337,000 acres, there is still plenty of room to live and play. There are more than 64,000 acres of federal and state land, more than any other county in the state. Reminiscent of the verdant English countryside from which its name was derived, the region offers a striking backdrop for every recreational pursuit, from snow skiing and ballooning to horseback riding and canoeing. Quiet colonial towns nestled among the horse farms, state parks, and lively amusement areas provide fascinating visual contrasts and a smorgasbord of leisure activities. In short, the possibilities for fun are endless.

Along the Kittatinny Mountains, Stokes State Forest stretches out over 15,328 acres of lovely mountain scenery providing a paradise for hikers (the Appalachian Trail and others are here), plus facilities for picnicking, boating, fishing, etc. High Point State Park is located in the extreme northwest corner of the state along the crest of the Kittatinnies. Consisting of 14,056 acres, it extends from the New York State line southwesterly for 8 miles, where it joins Stokes State Forest. A 220-foot obelisk dominates the summit of the highest mountain in New Jersey (elevation: 1,803 feet). For a nominal fee, visitors can take in a remarkable panoramic view of Pennsylvania's Poconos, New

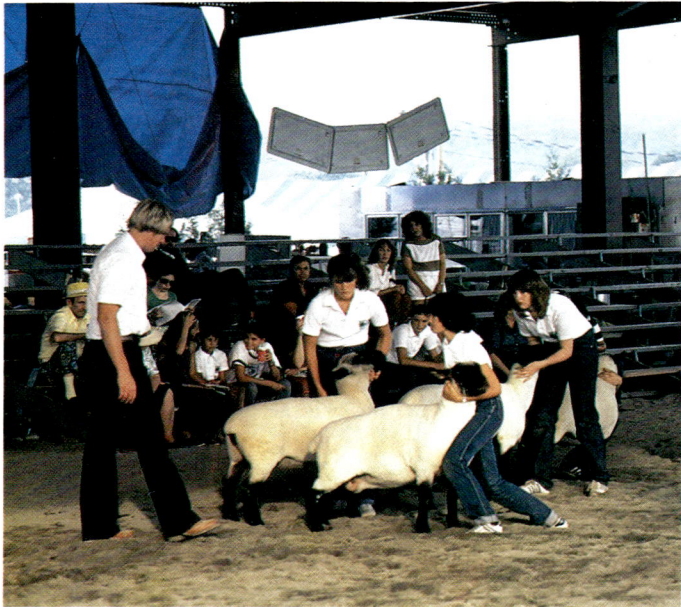

The Sussex County Farm and Horse Show is an annual event. Here, sheep are readied for judging. Photo by Michael Spozarsky

A lifesize schoolroom is one of several displays at the Clinton Historical Museum Village. Photo by David Greenfield

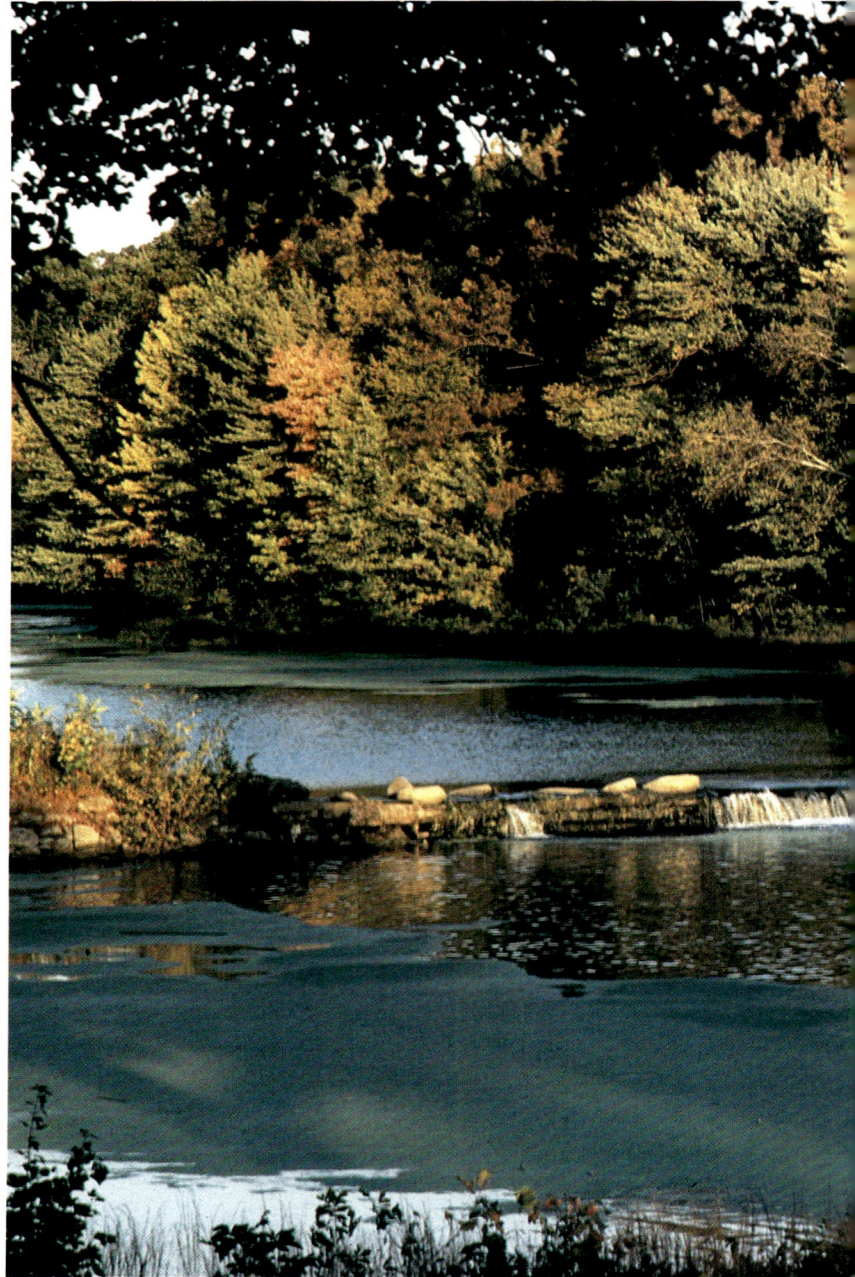

Autumn splendor surrounds Waterloo Village. Photo by Michael Spozarsky

the major ski areas in the region, both provide chair lifts, ski shops, restaurants, and lodges. Vernon Valley/Great Gorge even advertises "the world's largest snowmaking system."

One of the county's biggest attractions makes Vernon a scheduled stop on thousands of vacation itineraries each year. Action Park, billing itself as "the world's largest participating amusement park," features over 50 rides and attractions. The Space Farms Zoo and Museum, in Beemersville, will fascinate and educate visitors. Wildlife lovers have a chance to see indigenous New Jersey wildlife and exotic animals from around the world. The 100-acre family park features the world's largest bear: Goliath. Children will also delight in the Fairy Tale Forest in West Milford. Handcrafted life-size storybook characters jump from the pages of children's fables at the family storyland. Picatinny Arsenal

in Jefferson Township is a must for Revolutionary War buffs. The arsenal produced cannonballs and solid shot for the Continental Army. Olde Lafayette Village in Lafayette is one of the county's newest attractions. Featuring 50 shops, the village is patterned after Peddler's Village in Pennsylvania and Olde Mistick Village in Connecticut. Peters Valley Craftsmen is a resident craftsmen village within the Delaware Water Gap National Recreation area. The Old Monroe School House offers children the opportunity to see what school was like over 160 years ago. The tiny one-room structure, which is one of the few handhewn stone schoolhouses still in existence, is listed on both the State and National Register of Historic Places. Old books, hand slates, parkland. The park was named for Captain Anthony Swartwout, a British officer who, with his family, was slain by Indians during the French and Indian War. Hopatcong State Park, which is in the south-easternmost end of Sussex County, occupies 112 acres. The area has a beach on Lake Hopatcong (the state's largest lake) near Landing.

Port Morris and Stanhope share the distinction of being located at two of the 23 inclines of the famous Morris Canal, built in 1831. The canal, the engineering marvel of its day, was the first canal to take boats uphill via an inclined track. Between sea level at Newark and the Summit at Lake Hopatcong, each boat was raised 914 feet vertically within only 40 miles. Waterloo Village, which was once an important port on the Morris Canal, has been revived as a working restoration of the old community. The reconstructed village, with its Victorian and colonial homes, offers travelers a glimpse into what life must have been like along the banks of the old canal. Waterloo is also the home of the famous Waterloo Music Festival, held from May through October.

The Sussex County Historical Society Museum in Newton, the county seat, contains interesting items from Minisink and relics of colonial times. The Franklin Mineral Museum, in Franklin, was once the home of the world's largest zinc deposit. Its museum contains an exhibit of 200 kinds of minerals, many found nowhere else in the world. The museum also has a large exhibit of fluorescent minerals.

Antique fanciers can get their fill at several locations: the Lafayette Mill Antique Market, where 40 dealers offer items for sale; the Sussex County Antique Center in Branchville; and the many shops that line the main street in Andover, sometimes called "the antique town." However, antiques can be found throughout Sussex County, which is considered by many to be one of the best places in the East to find antiques.

Some restaurants in Sussex County: Black Forest Inn on Route 206, Stanhope; Perona Farms on Route 517, between Andover and Sparta; Samurai at 34 Lakeside Blvd., Hopatcong; Alpine Chalet in the Hidden Valley Ski Area, Breakneck Road, Vernon; The Walpack Inn on Route 615, Flatbrookville Road, Walpack Center; José Fitzgerald's Steakhouse and Lounge in the Americana Great Gorge Resort, Route 517, McAfee.

Lodging: American Great Gorge on Route 517, McAfee.

During winter months Lake Hopatcong is transformed into a virtual wonderland, attracting numerous ice-boating enthusiasts. Photo by Bob Krist

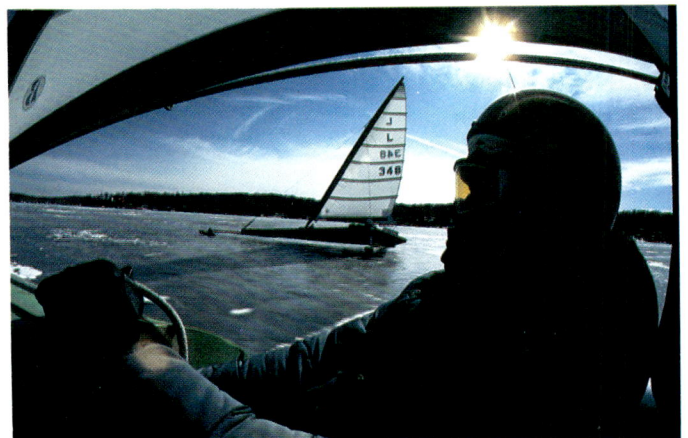

wooden desks, and a pot-belly stove all recapture the ambience of a by-gone era.

Frankford is the permanent site for one of the county's traditional events: the famous Sussex County Farm and Horse Show held in August. With everything from agricultural exhibits to one of the nation's top horse shows, the event is a true country fair. August is also the month in which the three-day Sussex Air Show takes place at the Sussex Airport.

Western fans of all ages can get a taste of the old West at Wild West City in Byram Township. The re-created Western town has 22 shows a day from May through October. The shows include reenactments of stagecoach holdups and even the gunfight at the OK Corral. Swartswood State Park, near Newton, is a wooded area surrounding a beautiful lake that occupies much of the 1,256 acres of

LOEWS GLENPOINTE HOTEL

Loews Glenpointe is a world-class hotel, featuring 345 spacious guest rooms and suites and fine restaurants in a magnificent 14-story tower that affords a commanding view of the New York skyline.

Loews Glenpointe is a world-class hotel, featuring 345 spacious guest rooms and suites and fine restaurants in a magnificent 14-story tower that affords a commanding view of the New York skyline.

The Loews Glenpointe restaurants and lounges offer guests accommodations that range from gourmet dining to casual elegance. Bronzini Ristorante, applauded by New York and New Jersey food critics, provides an elegant setting for some of Italy's and France's most classic and imaginative dishes. The Grill affords informal dining in a delightful garden-like terrace. The Glen Lounge, an exhilarating oasis, features a piano bar for guests' pleasure and convenience.

The fitness enthusiast is offered a myriad of options to either relax or stay in shape. A 20,000-square-foot health spa includes an indoor swimming pool, steam baths and saunas, an aerobic exercise facility, and a state-of-the-art physical fitness and cardiovascular center. Horseback riding, golf, tennis, and racquetball are within walking distance.

Sports fans can see soccer, NFL Giants football, hockey, and basketball at the nearby Meadowlands Sport Complex, or horse racing at the Meadowlands Racetrack. The Glenpointe serves as the headquarters hotel for the National Hockey League

The lush, tree-filled atrium of the Loews Glenpointe features abundant fresh flowers and comfortable furnishings in soft, contemporary tones.

All-Star Game and the Meadowlands Grand Prix Championship Auto Racing Teams.

Service is the hallmark of Loews Glenpointe. The Grand Prix Club offers a full floor of deluxe accommodations with a concierge to help guests with everything from theater tickets to limousines to foreign currency exchange. The hotel's amenities include on-premise parking, 24-hour room service and limousine service, a multilingual staff, and a shopping arcade offering a wide variety of items.

According to Loews Corporation president and co-chief executive officer Preston Robert Tisch, "We believe in responding to each guest's needs from the moment they enter the door to the moment they leave for their next destination."

Glenpointe's centralized location is ideally suited for the business traveler, vacationing family, or conventioneer. Glenpointe is only three miles from George Washington Bridge Crossing, minutes away from New York City, with direct access to major arteries I-95 and I-80, and easily accessible to Newark Airport.

The hotel features 21 meeting and banquet facilities, including a Grand Ballroom accommodating groups of up to 1,500 for business or social events. A professional conference service staff

This magnificent skylight and glittering chandelier reflect the sophisticated elegance of Loews Glenpointe.

Loews Glenpointe offers gourmet dining accommodations such as the Eronzini Ristorante, applauded by New York and New Jersey food critics.

assists meeting planners or social event planners in coordinating all the details necessary to make meetings, seminars, and social functions a success.

In addition to its convenient location and well-appointed facilities, the hotel offers the corporate meeting planner more than 20,000 square feet of conference and exhibition space, unlimited audiovisual capabilities, and hospitality and executive suites. The Grand Ballroom is capable of staging professional Broadway-style productions, elaborate trade shows, and exhibits—even car shows.

The hotel is part of the Glenpointe complex—a 50-acre mini-city including two office buildings and an elegant town house community. Alfred Sanzari is the owner, developer, and builder of the Loews Glenpointe Hotel and the Glenpointe Complex. The Glenpointe Centres East and West together comprise in excess of 500,000 square feet of commercial property. Alfred Sanzari's son, David, plays an integral role in the operation of Glenpointe and the family's other business interests.

Located in Teaneck, New Jersey, and opened in November 1983, the Loews Glenpointe is one of 13 hotels owned or managed worldwide by the Loews Corporation. Former U.S. Postmaster Preston Robert Tisch is president and co-chief executive officer of Loews Corporation, and Robert J. Hausman is chairman of the hotel division. Jona-

than Tisch, president of Loew's Hotels, was responsible for overseeing the completion and opening of Loews Glenpointe Hotel when he was vice-president/development.

Loews Corporation includes an insurance division, CNA Financial Corporation, Bulova Watches, Lorillard Tobacco Company, and a 24.9-percent investment in CBS Inc.

Preston Robert Tisch served for 20 years as chairman of the New York Convention and Visitors Bureau prior to his appointment as postmaster general. Now he is chairman emeritus of the New York Convention and Visitors Bureau. Efforts are under way by Loews New Jersey-based management to establish a Northern New Jersey Convention and Visitors Bureau to promote the area's outstanding business and recreational facilities.

A recipient of the New Jersey Business and Industry Association's Good Neighbor Award for its contribution to the area's economic growth, Loews Glenpointe's commitment to its community is apparent in its support of such organizations as the local chamber of commerce, the Boy Scouts of Bergen County, and the March of Dimes.

According to David Sanzari, "The healthy economic climate this area offers and its accessibility to the metro area have led to many *Fortune 500* companies locating their headquarters here. We look forward to its continued growth."

Pictured here is one of the 345 elegant guest rooms (and suites) that Loews Glenpointe offers its overnight visitors.

BERGEN COUNTY Beautiful Bergen County is situated in the extreme northeastern corner of the state. Its rivers are an important part of the region's picturesque and varied landscape. The massive cliffs of the Palisades, which stand like columns above the Hudson, and the steep slopes of the Ramapo Mountains in the northwest provide striking panoramas. The county's natural beauty, historical landmarks, and thriving commercial centers complement each other, making life in Bergen rich and stimulating.

The George Washington Bridge, one of the most famous structures in Bergen County, provides a link to New York City. Dedicated in 1931, the majestic drama of the steel span with its delicate mesh of cables makes it one of the most beautiful bridges in the world. Travelers coming into New Jersey via the bridge are let off near the Fort Lee Historic Park, operated by the Palisades Interstate Park Commission. Walkways and scenic overlooks provide commanding views of the Manhattan skyline, the lower Hudson River, and replicas of the Continental Army's 1776 fortifications. The visitor's center presents historic battles in a wide range of displays, including films, relics, mod-

els, etc. It was here that General Washington began his retreat after watching the fall of Fort Washington across the Hudson during some of the darkest days of the Revolution.

Another famous structure is situated in East Rutherford—The Meadowlands Sports Complex, a sports phenomenon. As one of the nation's most successful sports and entertainment facilities, "the Meadowlands" was the single most visited location in the early 1980s, drawing 10 million spectators each year. As one of New Jersey's popular attractions, the sports complex continues to prove itself. According to figures reported in the press, ticket sales at Giants Stadium alone totaled $26.8 million in 1986. Harness meets garnered $369 million in wagers, and thoroughbred races totaled $207.9 million during the same period. The complex includes the Meadowlands Race Track (home of the Hambletonian, the Kentucky Derby of harness racing) and Giants Stadium, built specifically for football. Both the New Jersey Nets (basketball) and the New Jersey Devils (hockey) play in the Brendan Byrne Arena.

Nearby, in Rutherford, the William Carlos Williams Center for

Opening day at the Meadowlands Racetrack on September 1, 1976, drew a crowd of 42,000 people. The success of the entire Meadowlands Sports Complex was inevitable. Photo by Bob Krist

the Arts draws visitors to its facility built around—and under—the Rivoli, a vaudeville theater built in 1922. Williams' home, near the center, is a historic monument. The Meadowlands Museum, the Railroad Station, and the Castle at Fairleigh Dickinson University, all in Rutherford, also provide interesting excursions. Only minutes away, Secaucus has one of the largest discount shopping areas in the country.

From the time of the first Dutch settlers up to the spectacular events of the nation's fight for freedom, history has dotted Bergen County's soil with remnants of distant times and events. Its history is New Jersey's and the nation's. A visit to Garretson Forge and Farm, in Fair Lawn, is an introduction to the rural lifestyle of one of Bergen's early Dutch families. Six generations of the Garretson family lived in Bergen from 1719 to 1950. The house represents a fine example of an important American architectural type—the stone homes built by settlers of Dutch cultural background. Its modifications and enlargements show the structure's evolution from an early farmhouse. The Campbell-Christie House at New Bridge Landing Historic Park in River Edge is another example of Dutch architecture. Built in 1774, its sturdy one-story sandstone walls and graceful gambrel roof survive as testimony to the county's unique Dutch cultural ancestry.

Other Dutch structures are also of particular note. The Old North Church in Dumont is listed in architectural publications as a perfect example of a Dutch Colonial building. The original building was constructed in 1735, but it was burned by Tories and rebuilt after the Revolution. The Steuben House, confiscated from its Loyalist owner, John Zabriskie, was presented to General Friedrich von Steuben in 1783 as a reward from New Jersey for his services during the war. Von Steuben never lived in the house and the owner eventually bought it back. The house is one of the few Dutch Colonial houses open to the public. In Ramsey, the Old Stone House Museum stands as a tribute to the efforts of a Dutch farmer named Ruloff Westervelt, who built it of rubble-stone with clay and chopped straw mortar. Three rooms are authentically furnished in the manner of the late 1700s. The Wortendyke Barn in Park Ridge is one of the few examples of the Dutch type. Although its exact construction date is unknown, its type was built in the 1800s.

Giants Stadium at the Meadowlands Sports Complex is the home of both the New York Giants and the Jets. Photo by Michael Spozarsky

SHERATON MEADOWLANDS HOTEL

The Sheraton Meadowlands' beautiful indoor swimming pool and deluxe health club enable guests to enjoy a workout, a session in the whirlpool or sauna, or a leisurely swim.

The 429-room, 20-story Sheraton Meadowlands Hotel, a center for business and pleasure travelers, offers a superb view of the Manhattan skyline and overlooks the Meadowlands Sports Complex.

New Jersey's Meadowlands area has undergone a miraculous metamorphosis. Once a largely overlooked and under-utilized stretch of tidal marshland situated in the shadows of the New York City skyline, the Meadowlands is today recognized as one of the country's leading economic and entertainment centers.

Home to a diverse corporate community and the world-renowned Meadowlands Sports Complex, thousands of business travelers and pleasure-seekers visit the Meadowlands area each year. Since its opening in the fall of 1986, the Sheraton Meadowlands Hotel has established a standard of excellence in accommodations, amenities, and service while welcoming visitors to the Meadowlands.

At the hotel's grand opening, New Jersey Governor Thomas H. Kean said the Meadowlands is "where the world is headed, and the Sheraton Meadowlands is likely where they'll stay the night." The 429-room, 20-story Sheraton Meadowlands Hotel has become a center for business and pleasure travelers, including many of the professional and collegiate athletes and performers appearing at the Meadowlands Sports Complex.

Certainly the site of the hotel has contributed a key ingredient to its success. Located on a 22-acre complex at the intersection of Route 3 and Exit 16W of the New Jersey Turnpike in East Rutherford, the Sheraton Meadowlands offers a superb view of the Manhattan skyline and overlooks the modern elegance of the Meadowlands Sports Complex, just across Route 3.

This central location makes the Sheraton Meadowlands easily accessible from midtown Manhattan via the Lincoln Tunnel, and from the Wall Street area via the Holland Tunnel. The proximity of Newark International Airport (12 miles away) and Teterboro Airport (three miles away) also makes the hotel convenient for travelers, as does easy access to New Jersey's major interstate highway network. The Sheraton Meadowlands adds to guests' convenience by providing shuttle transportation between the hotel and airports, as well as offering the option of public transportation both in and out of New York City on a regularly scheduled basis.

Hotel guests capitalize on the Sheraton Meadowlands' location to experience the pleasures of the area's world-class entertainment options, including New York theater, dining, and art; the

The Club Lounge Bar contributes to the overall ambience of casual comfort and luxury.

excitement of thundering hooves at Meadowlands Racetrack; spectacular shopping at the Meadowlands' many designer outlets; and a wide range of other activities within minutes of the hotel's front doors.

The hotel's location, just a "long pass" from the Sports Complex, home of the National Football League Giants and Jets, the National Basketball Association Nets, and the National Hockey League Devils, has had a tremendous impact on the Sheraton Meadowlands' ambience and decor. The Meadowlands Racetrack, home of some of the finest thoroughbred and standardbred horse racing in the country, has also contributed substantially to the hotel's distinctive style.

That style can best be expressed as casual elegance. Throughout the hotel, guests encounter an attention to quality and detail that makes the Sheraton Meadowlands Hotel unique among its peers. The excitement of the Sports Complex and the vibrancy of the thriving Meadowlands business community add to the feeling of energy at the Sheraton Meadowlands Hotel.

Dining at the hotel's signature restaurant, Roses, one experiences the clublike feel of dark wood paneling, original oil paintings, and leather, while sampling its traditional grillroom-style fare. The light and airy café-like setting of the Terrace Restaurant provides a bright and sunny counterpoint perfect for brunch, lunch, or anytime.

Even the Sheraton Meadowlands' deluxe health club and indoor swimming pool contribute to the hotel's distinct character, providing a stunning view of the New York skyline from its atrium-like setting.

Recognizing the potential of the Meadowlands area, the Metropolitan Life Insurance Company chose the site for the Sheraton Meadowlands and the Metropolitan Executive Towers directly across from the Meadowlands Sports Complex. The development of a world-class hotel and office complex represent

a commitment to excellence—a commitment shared by Metropolitan Life and the Sheraton Corporation.

No element of the Sheraton Meadowlands more aptly demonstrates this commitment than the hotel's conference center. Conveniently laid out on one floor and strategically planned to provide state-of-the-art audio and visual capabilities, the Sheraton Meadowlands' conference facilities are meeting the needs of a wide variety of groups with graceful, stylish settings for memorable occasions of all kinds and sizes.

The Sheraton Meadowlands also offers a seasoned convention and catering staff, as well as a professional coordinator to organize meetings and social affairs. One of the biggest advantages offered by the Sheraton Meadowlands in terms of both social and business function is, literally, the size of its Diamond Court Ballroom and adjacent Derby Ballroom. These facilities constitute the largest hotel ballroom complex in northern New Jersey, capable of accom-

modating groups of up to 2,000 people.

Convenience is expressed in additional ways at the Sheraton Meadowlands, including free parking adjacent to the hotel for 450 cars, as well as a five-story parking facility with space for an additional 1,000 cars. A lobby retail area where guests can shop for convenience items and luxury gifts contributes to the hotel's accommodating atmosphere.

The employees of the Sheraton Meadowlands also contribute significantly to the world-class performance standard of the hotel. The facility has been highly ranked among Sheraton's corporate and resort hotels throughout North America in terms of overall guest satisfaction.

Perhaps this attention to the needs and desires of guests at the Sheraton Meadowlands Hotel has contributed more to its overwhelming success among business and pleasure travelers than the hotel's fine facilities, amenities, and location combined.

THE WOODCLIFF LAKE HILTON

The Pearl River Hilton in Rockland County, the sister hotel of The Woodcliff Lake Hilton, maintains the ambience of a French château.

In the past, country inns served the needs of travelers and provided the setting for local social activities. Those concerns are now served by one of the most prestigious hotels in New Jersey, The Woodcliff Lake Hilton—the first in the state awarded four stars by Mobil and four diamonds by the Automobile Association of America.

There traditional care and hospitality are found in the midst of the most modern comforts and conveniences, featuring such amenities as luxurious room accommodations; gourmet dining; complete recreational facilities, including three pools, tennis, racquetball, exercise room, sauna, whirlpool, and jogging trail; and a variety of support services for business needs.

One would think that these accomplishments are the result of long experience in the hospitality industry, but that is not the case. The Woodcliff Lake Hilton rose out of the firm belief of the owners, the Maloney family, in quality service and care that had first manifested itself in the development of superior nursing homes in Bergen County.

The Maloney family, recognizing the need for health care for an expanding senior population, opened their first facility in Paramus in 1956. This establishment was followed by Valley Nursing Home in Westwood in 1960, and later the Pine Rest and Cupola locations in Paramus. The Maloney reputation for quality care and service grew as well.

In 1978 the Maloneys, joined by son William Jr., set out to diversify. Realizing the potential for corporate development in the areas adjacent to the Garden State Parkway, they purchased a 21-acre site from the Tice family farm for the first luxury hotel in Bergen County.

The Woodcliff Lake Hilton opened on April 1, 1980, with 215 rooms. It was expanded to 350 rooms in 1983 to meet the burgeoning lodging and meeting demands of the hotel's multiplying corporate neighbors, such as BMW, Timeplex, A&P, and others.

The family expanded again with the opening of its Pearl River Hilton in nearby Rockland County. Where The Woodcliff Lake Hilton has the air of a country inn, its sister hotel maintains the ambience of a French château—strategically located adjacent to the Blue Hills Golf Club.

In tribute to their commitment to the highest standards of service and quality, the Maloney family has been awarded a contract to build the first new oceanfront hotel in decades, the Ocean Place Hilton in Long Branch, New Jersey.

The Maloneys employ staffs of more than 1,000 people among their hotel and health care facilities. Their success is evident because there is a bottom-line performance that is paramount: the commitment of each staff member to offer the finest in hospitality, care, and service to each and every guest.

The Solarium Lounge at The Woodcliff Lake Hilton.

The hillsides of Warren County pro-
vide a scenic backdrop for these
hang gliders. Up, up, and away!
Photo by Michael Spozarsky

Bergen County is unique in the abundance, variety, and architectural quality of its early stone houses, although adjacent areas of New Jersey and New York have similar edifices. In the Survey of Early Stone Houses of Bergen County, conducted in 1978-1979, 230 houses were identified and recorded. Of those, 208 retained sufficient architectural integrity to be placed as a thematic group on the Register of Historic Places in 1980. Most of these were also listed on the National Register of Historic Places in 1983, 1984, and 1985. These survivors of the distant past represent tangible evidence of Bergen County's participation in the formative years of the county, state, and nation.

Hackensack, the Bergen County seat, was built around a New England-style green, even though its central structure was the red sandstone First Dutch Reformed Church built in 1696. The large Bergen County Courthouse is a domed, neo-classical edifice. The USS *Ling,* a remnant of more recent history, is on display in Hackensack. An example of the last of the fleet-type submarines that patrolled our shores during World War II, the *Ling* was stricken from the Navy Register in December 1971 and donated for preservation as a memorial.

Along the Hackensack River, an idyllic setting memorializes a bloody event that took place during the Revolution. In 1778, British forces attacked 100 sleeping members of the Third Dragoons of Virginia led by George Baylor, killing 67. The Baylor Massacre Burial Site commemorates those patriots. Gethsemane Cemetery in Little Ferry played a role in the passage of New Jersey's civil rights legislation. The controversy surrounding the burial of Samuel Bass led to the passage of the "Negro Burial Bill" in March 1884. It is one of three civil rights bills enacted in New Jersey in the 1880s. The Camp Merritt Memorial Monument in Cresskill is an obelisk patterned after the Washington Monument; it commemorates a World War I camp. Washington Spring Garden in Paramus, which is part of Van Saun Park, lures tourists with its beauty and tradition; George Washington himself drank from its clear waters. The event took place on September 13, 1780, while General Washington's large encampment spread out from present-day Route 4 into Emerson.

The Pascack Historical Society Museum in Park Ridge houses one of the most colorful examples of Early American ingenuity: the world's only wampum drilling machine, which made wampum for trade with the Indians. The machine was invented by two of John Campbell's sons. The museum also has a general store, colonial kitchen, nineteenth-century parlor, and other displays depicting early American life in the Pascack Valley.

This opulent sitting parlor is just one of 78 rooms in the Ringwood Mansion. Photo by Mark Gibson

Custom woodwork is extensive throughout Ringwood mansion. Photo by Mark Gibson

Ringwood Manor, presently a state park, had been privately owned between 1810 and 1930 by leading families of the local iron industry. Photo by David Greenfield

Railroad buffs will delight in the memorabilia to be found at the Old Station Museum and Caboose and Wanamaker Building in Mahwah. The Hermitage, in Ho-Ho-Kus, survives as a prime example of the Gothic Revival style, and is one of the few remaining works of architect William Ranlett. The eighteenth-century house was the residence of Theodosia Provost, later the wife of Aaron Burr. Ringwood Manor is another example of a house with a colorful past. In the eighteenth and nineteenth centuries Ringwood was an important iron-producing center, and Ringwood Manor was home for its famous ironmasters. The forges supplied a good percentage of America's munitions for the Revolution. Washington was a frequent visitor when his map maker, Robert Erskine, was ironmaster. Partially destroyed by the British in 1765, Robert Ryerson rebuilt it in 1810. From then until 1930, the 78-room mansion was occupied by Ryersons, Coopers, and Hewitts, the leading iron families. They lived in luxury, and the valuable collection of furnishings and Americana they amassed is on display at the manor.

Those interested in aviation history will be fascinated by the Aviation Hall of Fame at Teterboro Airport, which contains artifacts from the early days of flying.

Bergen County's recreational options do not end with its historical landmarks and museums. Much of the county is set aside for the enjoyment of nature and outdoor activities. Its park system caters to every resident and tourist regardless of personal taste or means. One can pursue every outdoor endeavor, including hiking on wilderness trails, fishing, feeding ducks, cycling on paths next to streams and ponds, camping, visiting a wildlife center, taking in the zoo, or a whole panoply of similar pastimes. And one can enjoy these activities only minutes from thriving, sophisticated suburban and urban areas. With most parks open from 8 A.M., enthusiasts can get an early start on relaxing and having fun.

Some restaurants in Bergen County: Villa Cesare, Kinderkamack Road at Lincoln Avenue, Hillsdale; La Petite Auberge, 44 East Madison Avenue, Cresskill; Bronzini at Loew's Glenpointe Hotel, 100 Frank W. Burr Boulevard, Teaneck; Il Villagio at 651 Route 17 N, Carlstadt; Pegasus, atop the Meadowlands Racetrack, Meadowlands Complex, East Rutherford; Il Calcio at 12 Tappan Road, Harrington Park; La Dolce Vita at 316 Valley Brook Avenue, Lyndhurst; Wyckoff Inn at 179 Godwin Avenue, Wyckoff; Captain's Table at 644 Pascack Road, Westwood, Washington Township; Arthur's Sheraton Heights Hotel on

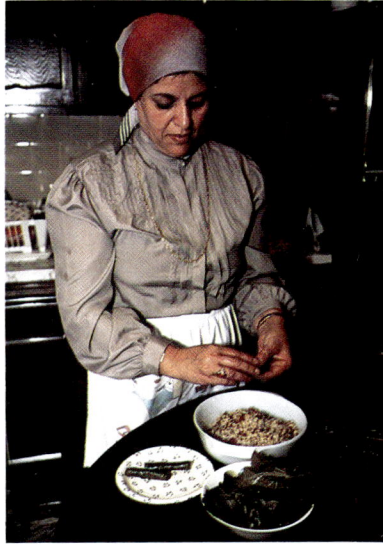

Stuffed grape leaves are a specialty of this Palestinean woman in North Bergen. Photo by Carol Kitman

A delightful display of Japanese delicacies were photographed with their proprietor in Fort Lee. Photo by Bob Krist

Polifly Road, Hasbrouck Heights.

Lodging: Loew's Glenpointe Hotel at 100 Frank W. Burr Boulevard, Teaneck; Sheraton Meadowlands on Sheraton Plaza Drive, 2 Meadowlands Parkway, East Rutherford; Sheraton Heights at 650 Terrace Avenue, Hasbrouck Heights; Treadway Saddle Brook Hotel on the Garden State Parkway at I-80, Saddle Brook; Best Western Oritani Motor Hotel at Route 4 and 414, Hackensack Avenue, Hackensack; Ramada Inn at 100 Chestnut Ridge, Montvale; Saddle Brook Marriott on the Garden State Parkway and I-80, 1 Saddle Brook; Sheraton International Crossroads at 1 International Boulevard, Mahwah; Woodcliff Lake Hilton at 200 Tice Boulevard, Woodcliff Lake; Holiday Inn-Fort Lee at 2117 Route 4 East, Fort Lee; Quality Inn at 283 Route 17, Hasbrouck Heights.

PASSAIC COUNTY Paterson's unique history as the first planned industrial city makes it the perfect stop for those interested in investigating America's industrial beginnings. The Great Falls of the Passaic River, where the water plunges 70 feet into a vertically walled chasm, is as impressive today as it was when Alexander Hamilton viewed it. The mammoth waterfall, second in the East only to Niagara Falls, pro-

vided energy for the mills that grew up nearby. A three-tiered raceway, built by Pierre L'Enfant, brought the water to the mills.

By the 1820s, Paterson was known as the Cotton City. The locomotive industry began in 1837. Eventually, the cotton mills converted to silk when English silk workers flocked to Paterson after a change in tariffs. Other immigrants followed as the city grew into an industrial center. Today, a statue of Hamilton, which stands in Great Falls Park, peers out over the town which evolved from his idea to build an industrially independent America. The city's old mills, a treasure trove for industrial archaeologists, are now part of a National Historic District. Plans have been made to convert many of the mills to contemporary uses.

On the other side of the Passaic River, the *Finian Ram,* one of the first modern submarines, is on display in Westside Park. The submarine was built by John P. Holland in 1880. Another Holland submarine may be seen in the Paterson City Museum. The 14-foot submarine was built in 1878 in Samuel Colt's Paterson gun mill, birthplace of the first revolving rifle. Both inventors met with difficulty: Holland's submarine sank, and Colt couldn't persuade the government to buy his rifles until the Mexican War.

Built in 1891 for silkmill owner Catholina Lambert, Lambert Castle is now the current residence of the Passaic County Historical Society. Photo by Michael Spozarsky

The Lambert Castle, a mansion that resembles an old English castle, was built in 1891 by silk-mill owner Catholina Lambert, who came to the United States as a steerage passenger. His success as a silk manufacturer allowed him to become an avid art collector, and he used the castle to store his vast collection, worth in excess of $1 million. In 1925, the castle became part of what is now the Garret Mountain Reservation. The castle provides an unparalleled view of Paterson, the sprawling Passaic Valley, the Ramapo and Kittatinny mountains, and the New York skyline. The Passaic County Historical Society occupies the ground floor; its exhibits include magnificent oil paintings, antique furniture, a stunning silver spoon collection, and antique items relating to the silk industry. Another mansion can be found in Wayne. George Washington made his headquarters at the home of Theunis Dey during July 1780. Later, Washington moved his troops up the Hudson River Valley. Then, after the discovery of Benedict Arnold's treachery, Washington moved back to New Jersey to guard himself against a kidnapping attempt. During October and November 1780, he again made the Dey Mansion his home. The mansion is one of the state's finest colonial houses. It has been restored and is furnished with eighteenth-century pieces. The mansion is part of the Passaic park system.

Besides Passaic County's unique historical spots, the county has outstanding outdoor areas. Over 40 lakes and ponds, nine reservoirs, and five country clubs contribute to the excellent fresh air activities available to residents and tourists alike. The numerous forests and parks provide facilities for hunting, fishing, hiking, bird watching, cross country skiing, swimming, and boating. The county park system contains five parks, the largest of which is Garret Mountain with 569 acres. Large, undeveloped tracts of land lie within Abraham S. Hewitt and Norvin state forests, Ringwood State Park, Wanaque Fish and Wildlife Management Area, and Newark Watershed Conservation District. Boating and fishing are available at Greenwood Lake.

Some restaurants in Passaic County: Roberto's of Clifton at 101 Route 46W, Clifton; Bartolo's at 331 Union Boulevard, Totowa; Bel'Vedere at 247 Piaget Avenue, Clifton; L'Auberge De France at 2320 Hamburg Turnpike, Wayne; La Riviera at 421-27 Piaget Avenue and Route 46, Clifton; Sukeroku at 68 Route 23, Little Falls.

Lush tree growth abuts a freshly tilled field in northern New Jersey. Photo by Rich Zila

Lodging: Ramada Inn at 265 N.J. 3, just west of N.J. 21, Clifton; Regency House Motor Inn at 140 Highway 23, Pompton Plains.

MORRIS COUNTY Morris County, like most of the northern portion of the state, is endowed with a richly textured past. Its central position in the north made it easily defensible, and Morristown became the "Military Capitol of the American Revolution." Morris County's status as one of the great historical counties largely stems from the vital role it played in the nation's fight for freedom. Cannon and shot were produced in a number of furnaces in the area, and powder was manufactured there for the army. Washington and his troops wintered near Morristown after the victorious battles of Princeton and Trenton in 1777, and again two years later. During these tumultuous years, Morris County saw history unfold. Benedict Arnold was tried for treason in Morris County, and the young Colonel Alexander Hamilton courted Betsy Schuyler, who lived just around the corner from Washington's headquarters at the Ford Mansion.

But Morris County's history doesn't end with the nation's first war. Its great iron industry flourished and became the nucleus of its econ-

Lake Hopatcong proves to be fruitful fishing grounds, all year long! Photo by Michael Spozarsky

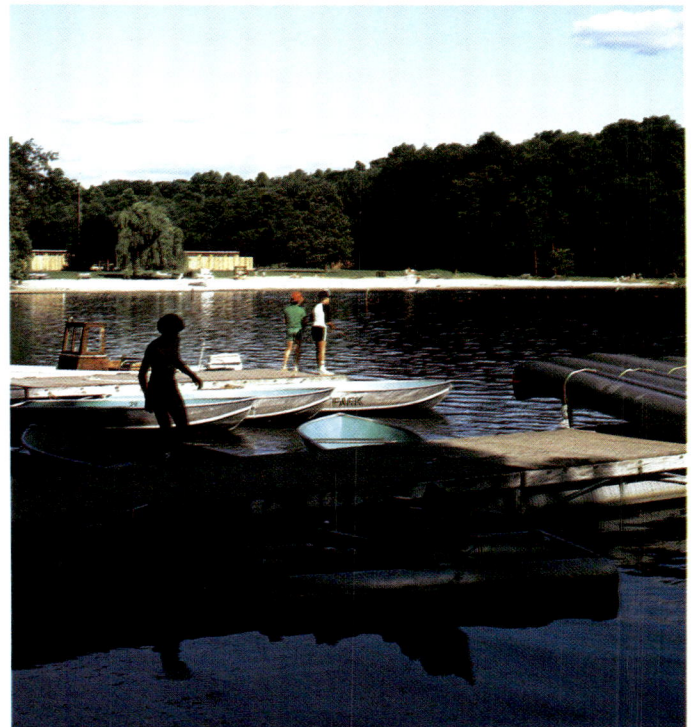

Swartzwood State Park is a fine site for boating. Photo by Michael Spozarsky

omy in the nineteenth and twentieth centuries. Through the mines, the area attracted an abundance of allied industries: mine equipment plants, nail production, steel foundries, and a giant powder company. The now famous Morris Canal was conceived by George Maculloch of Morristown. The advent of the railroad brought men and women of wealth to the beauteous landscape. Before long, Morristown was transformed into a residential haven for the wealthy who sought seclusion in its untouched woodlands and fields. Notables from the world of finance, Wall Street, art, and literature came to dwell in the area. The posh suburban retreat numbered among its residents Bret Harte, cartoonist Thomas Nast, and inventor Alfred Vail. It was frequently said that "within a mile of the Morristown Green are located the homes of more captains of commerce than anywhere in the world."

It is not surprising that Morris County's vivid past draws so many visitors each year. Its ample outdoor recreational areas and large land preserves add to its allure. The Morris County Park System, which encompasses more than 8,000 acres of recreational and educational facilities, operates 23 sites, including three golf courses, two arboreta, and the Mennen Sports Arena. Its aggressive education program has benefited families and school excursion groups alike. Visitors can delight in a firsthand view of how grain was processed in the 1880s by visiting Cooper Mill, the oldest existing mill in Morris County. Fosterfield's, a living historical farm, provides a look back to farming

as it was done in the 1880s. The George Griswold Frelinghuysen Arboretum is of particular note. The seasonal floral displays include cherries, magnolias, azaleas, roses, annuals, and bulbs. Its special features include a self-guiding trail, and a braille nature trail. Its Administration Building, on the National Register of Historic Places, contains a fine horticultural and botanical library.

The Great Swamp was established in 1960. Swamp woodland, hardwood ridges, cattail marsh, and grassland typify this 6,818-acre refuge. The swamp contains many large old oak and beech trees, stands of mountain laurel, and species of other plants of both northern and southern botanical zones. There are over 223 species of birds in the confines of the swamp area. Mammals that live in the refuge

include the white-tailed deer, beaver, muskrat, raccoon, skunk, red and grey fox, woodchuck, and cottontail rabbit. There are also fish, reptiles, and amphibians, including the rare and endangered bog turtle and the blue-spotted salamander.

Morris County's Revolutionary War sites are under the administration of the National Park Service. Among the most colorful is the Ford Mansion, which was Washington's headquarters during his second winter here. Some of its original furniture remains, including the General's writing desk where he penned anxious letters during those days. The Ford Mansion became the military capital of the nation for nearly seven months. The Historical Museum and Library are behind the house, where you can examine items relating to the period. Fort Nonsense, built in the spring of 1777, got its name because it is said that Washington had his men build it to help them avoid depression, and to prevent indolence and desertion. At the Grand Parade, troops were drilled daily. There are reconstructed huts at the site of the encampment of the Pennsylvania Line. "Living history" displays take place here in the summer. Jockey Hollow was the encampment for the rest of Washington's troops. An eighteenth-century farmhouse served as the headquarters of General Arthur St. Clair, commander of the Pennsylvania Line. Owned by Henry Wick, the house, like so many of the era, was put to use in the war effort. Charmingly restored, the house is a good example of the Cape Cod style.

Morristown is also the home of the Morris Museum of Arts and Sciences. This excellent museum is a treasure trove of interesting displays: dinosaur bones, railroad models, minerals, kachina dolls, puppet shows, Indian costumes, and push-button displays of physical principles. The Schuyler-Hamilton House is of interest because of its eighteenth-century furnishings and its background. Betsy Schuyler, the future wife of Alexander Hamilton, lived in the house with her uncle, Surgeon General Dr. John Cochrane.

Madison is known for its colleges: there are three within its limits. Each institution has its unique physical features and buildings. Drew University was founded in 1866 by stock speculator Daniel Drew. Its main building, Mead Hall, was built in the Southern plantation style by William Gibbons. The university's Rose Memorial Library contains rare publications on John Wesley and early Methodism. Fairleigh Dickinson University's Florham-Madison campus was built around the Stanford White Mansion of Mrs. Florence Vanderbilt Twombly, granddaughter of Cornelius Vanderbilt. The college of St. Elizabeth, in Convent Station, has a charming Shakespeare Garden. The Museum of Trades and Crafts can also be found in Madison at the Willis James Memorial Library.

Historical markers are in plentitude along every byway in the county, making Morris' historical past among the richest in the state.

Some restaurants in Morris County: Hunan Taste at 67 Bloomfield Avenue, Denville; The Black Orchid in the Headquarters Plaza Hotel, Morristown; The Publick House at 111 Main Street, Chester; The Tarragon Tree at 225 Main Street, Chatham; The Grand Cafe at 42 Washington Street, Morristown; Calaloo Cafe at 190 South Street,

Morristown; Il Cappriccio at 633 Route 10, Whippany; Joie De Vivre at 72 Madison Avenue, Madison; Rod's 1890s Ranch House in The Madison Hotel, Route 24, Convent Station; Larison's Turkey Farm Inn at Routes 206 and 24, Chester; Le Délice at 302 Whippany Road, Whippany; Silver Spring Farm on Drakestown Road, Flanders.

Lodging: The Madison Hotel at 1 Convent Road, Convent Station; The Publick House at 111 Main Street, Chester; The Governor Morris Inn at 2 Whippany Road, Morristown; Parsippany Hilton at 1 Hilton Court, Parsippany.

This waiter, serving dinner at a Spanish-Portuguese restaurant in Harrison, takes his work very seriously. Photo by Carol Kitman

The Great Swamp National Wildlife Refuge provides visitors with a platform walkway, thus maximizing appreciation while minimizing discomfort. Photo by David Greenfield

Sacred Heart Cathedral rises above the newly blossomed cherry trees of Branch Brook Park in Newark. Photo by David Greenfield

ESSEX COUNTY Essex County's claim to fame has always rested upon Newark's illustrious history. As the largest city in New Jersey, Newark is still one of the northeast's main transportation hubs. Its international airport and seaport, highways, daily commuter trains, bus lines, the most advanced freight-transporting facilities in the United States, and its own subway make the city a modern transportation wonder. However, its past is worth exploring.

Newark's "Four Corners," at the intersection of Broad and Market streets, has been the city's core since its beginning. The First Presbyterian Church, or Old First Church, completed in 1797, is one of Newark's many distinctive houses of worship. Trinity Cathedral in Military Park was built in 1747. The park itself got its name because it served as a training site for Continental troops. It was here that Thomas Paine declared that "these are the days that try men's souls."

Newark's unusual buildings make it a history and art enthusiast's dream. The Art Deco style of Penn Station, Newark's main terminal, makes the building unique. There are also others of interest: the North Reformed Church, a Gothic structure built in the nineteenth century;

the Ballantine House, a restored Victorian mansion; and the Art Deco Bell Telephone Company building. The Essex County Courthouse, designed by Cass Gilbert, has a statue of Lincoln in front that shows the former president in a rare, informal pose. The statue is by Gutzon Borglum.

Newark's county park system, the oldest in the United States, has helped to retain beautiful areas that still offer a respite from hectic lives. One of the loveliest of these, and the first in the park system, is Branch Brook Park. Dedicated in 1895, the park's 360 acres were designed by Frederick Olmstead. Its 2,500 oriental cherry trees flower each April, creating a natural spectacle that is larger and more spectacular than the one found in Washington, D.C. Nearby, Sacred Heart Cathedral, styled after Rheims Cathedral, houses a world-famous organ. Sacred Heart is one of the largest cathedrals in America.

History, science, and artworks are housed in the Newark Museum, a three-story building. Newark's oldest schoolhouse (1784) is located in the sculpture garden behind the structure. The Newark Fire Museum is also in the garden. It contains antique firefighting

The work space and library of an extraordinary man, Thomas Alva Edison, is pictured here. Photo by David Greenfield

The conservatory of Glenmont, the 20-room home of Thomas Edison, is maintained with its original decor. Photo by David Greenfield

Tourists may visit the West Orange Laboratory of Thomas Edison where various inventions remain on permanent display. Photo by David Greenfield

equipment. The Ballantine House, entered through the museum, contains five rooms restored in high Victorian style. The New Jersey Historical Society, also in Newark, is devoted to the preservation and presentation of New Jersey and American History. Exhibits include transportation and industry, household utensils, textiles, decorative arts, jewelry and Indian arts, as well as painting, drawings, and prints. There are views of 567 New Jersey towns before 1875, the first map of the state and its charter signed by King Charles II in 1664, a portrait of General George B. McClellan of Civil War fame, and equipment invented and perfected by Thomas Edison in West Orange from 1890 to 1920.

In his West Orange laboratory, Thomas Alva Edison, America's most famous and prolific inventor, created more than half of his 1,093 patented inventions. The work at West Orange set the standard for all the industrial research labs that followed. The many interesting exhibits to see in West Orange include a replica of the Black Maria, the first motion-picture screening room, and Glenmont, Edison's home. Its 20 rooms are still furnished as they were when he was alive. The Edison

National Historic Site marked the 100th anniversary of Edison's West Orange laboratories this year. The home of another great American can be found in Caldwell. The Grover Cleveland Birthplace preserves the residence of our twenty-second and twenty-fourth president. Cleveland was the only president to win after losing a second-term election. He is also the only president from New Jersey.

Southwest of Orange, more cherry trees can be found in South Mountain Reservation, which has ice skating year round. Close by, the Turtleback Zoo has more than 750 animals and amusements for children.

In Montclair, the Montclair Art Museum, founded in 1912, is a neoclassic building set in a wooded area; more than 50 varieties of trees are identified and labeled. Besides its collection of American art, the museum presents art and artifacts of the seven major North American Indian cultural groups. Other collections include textiles, silver, ceramics, decorative arts, European paintings, Japanese prints, and one of the finest collections of Chinese snuff bottles in America.

All in all, Essex County is a place where past and present live

Numerous ethnic festivals are held every year at Liberty State Park in Jersey City. New York City provides a stunning backdrop for such occasions. Photo by David Greenfield

amiably in close proximity. Its parks system, a forerunner of others in the country, maintains 5,645 acres of beautiful parks, reservations, and golf courses amidst a highly populated and active commercial area.

Some restaurants in Essex County: The Manor at 111 Prospect Avenue, West Orange; Nanina's In The Park at 540 Mill Street, Belleville; Pal's Cabin on Prospect and Eagle Rock avenues, West Orange; Il Tulipano at 1131 Pompton Avenue, Cedar Grove; Gitane at 52 Vose Avenue, South Orange; Spain, 419 Market Street, Newark.

Lodging: Holiday Inn at 550 W. Mt. Pleasant Avenue, Livingston; Hilton Gateway at the Gateway Center, Raymond Boulevard, Newark; Holiday Inn-International Airport North at 160 Holiday Plaza; Holiday Inn-Jetport at 1000 Spring Street, Elizabeth; Howard Johnson's Hotel, Route 1 South and Haynes Avenue, at the Airport; Sheraton Inn-Newark Airport at 901 Spring Street, Elizabeth; Marriott Hotel-Newark Airport at Newark International Airport; Sheraton Fairfield Hotel at 689 U.S. 46 East, Fairfield.

HUDSON COUNTY Hudson County is often called the "gateway" to New York. Its strategic location in the heart of the largest urban complex in the nation commands an astonishing across-the-Hudson view of Manhattan's spectacular skyline. Newark Bay and the Passaic River are to its west, while the Kill Van Kull separates it from State Island on the south. The county is divided by the Hackensack River and the famous Meadowlands. As one of the most populous counties in the state, the region is a major manufacturer, with an 8.5 percent share of the state's total. Its proximity to Port Newark and the three major metropolitan airports puts it at the center of world trade.

Hudson County's three major cities provide the avid tourist with a great many diverse activities. Whether you are interested in tracking historical events or enjoying the fabulous new waterfront developments and Liberty State Park, Hudson County provides striking contrasts and fabulous scenery.

Taking either the Lincoln or Holland tunnels, the visitor from

A world-famous symbol, the Statue of Liberty, stands before Liberty State Park, northern New Jersey, and beyond. Photo by Michael Spozarsky

New York City emerges in Hudson County. The Lincoln Tunnel funnels people into Weehawken, where one of the most famous duels in history took place in 1804. Alexander Hamilton, a Federalist, and Aaron Burr, who believed more strongly in states' rights, had quarrelled for years over the matter. When Hamilton went public with his criticism, Burr challenged him. Hamilton, firing his pistol in the air, was killed. Oddly, Hamilton's son had lost his life in a duel three years earlier in the same spot. His career ruined, the Vice President fled indictment for murder by escaping to the South. A commemorative plaque decorates the spot.

The Holland Tunnel feeds into Jersey City, a sea terminus and an industrial center. Its landmark is the Colgate Soap factory with its famous huge clock. The Pulaski Skyway connects Newark to the Holland Tunnel. To the south, the strikingly beautiful Bayonne bridge, the longest steel-arch bridge in the world, spans the Kill Van Kull River to connect Bayonne to Staten Island.

Liberty State Park, an 800-acre park, is the most frequently visited state park in New Jersey. Its outstanding view of New York and the ferry service to Ellis Island and the Statue of Liberty ensure its status as a prime tourist attraction. Commemoration of Miss Liberty's 100th birthday on July 3-4, 1986, brought thousands of people to the park to celebrate the national event. The park, which is now under development, will have an 18-hole championship golf course, a science museum, and many other amenities. The park is part of the sprawling and aggressive rejuvenation of the New Jersey Hudson River waterfront sometimes called the Gold Coast, an 18-mile strip of land extending from the George Washington Bridge on the north to Bayonne on the south.

Hoboken is another area that has gone through a transformation. With factories and warehouses being turned into quaint townhouses and shopping areas, the town has entered an exciting renaissance. Its readiness to participate in the area's future is indicative of its longstanding vitality. Hoboken's illustrious past saw many firsts: the first steamboat, the first yacht club, the first organized baseball game (between the Hoboken Knickerbocker Giants and New York), and America's first "legitimate" teen idol—Frank Sinatra. The heirs of Hoboken's founder, Colonel John Stevens, founded Stevens Institute of Technology in 1870, the first college of mechanical engineering in America.

The recent restoration of the Jersey City Train and Ferry Station at Liberty State Park yielded outstanding results. Photo by Michael Spozarsky

Locals celebrate the occasion of this Railroad Festival at the Hoboken train terminal. Photo by David Greenfield

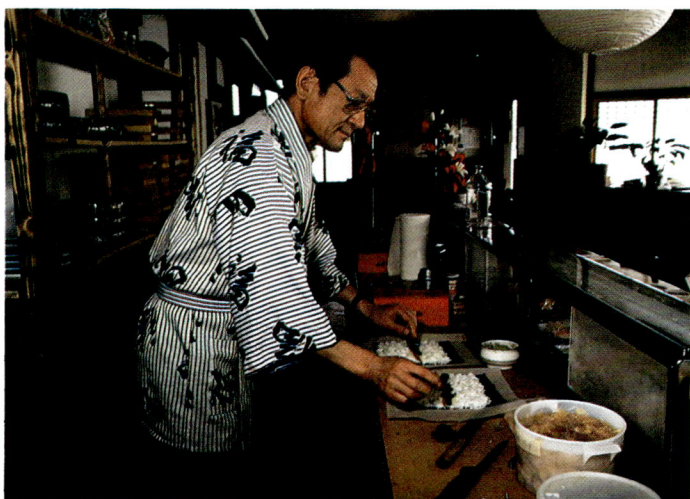

Eel rolls are masterfully prepared at this Japanese-Korean restaurant in Jersey City. Photo by Carol Kitman

Some restaurants in Hudson County: Shanghai Red's on Pier D-T, Lincoln Harbor, Weehawken; Aratusa Supper Club at 1 Meadowlands Parkway, Secaucus; Brazilian-Portuguese Pavilion at 10-12 N. Third Street, Harrison; Clam Broth House II at 540 Fifty-fifth Street, West New York; Laico's at 67 Terhune Avenue, Jersey City; Hunan Dynasty at 80 River Street, Hoboken; The Brass Rail at 135 Washington Street, Hoboken; Helmer's at 1036 Washington Street, Hoboken.

Lodging: The Meadowlands Hilton at 2 Harmon Plaza, 1 mile southeast of Meadowlands Stadium; The Hilton at Harmon Meadows, 300 Plaza Drive, three-quarters of a mile southeast of Meadowlands Stadium.

TOURISM New Jersey's considerable reputation as a tourist destination has made tourism the state's largest industry. According to figures from New Jersey's Division of Travel and Tourism, the industry provides employment for more than 218,900 people. According to the U.S. Travel Data Center, New Jersey surpassed the rest of the country in tourism growth: an 18 percent hike in 1983 ballooned to 26.5 percent in 1984. Revenues of $13.1 billion flowed into the state in 1985. In calendar year 1986, 54 million visitors came to New Jersey, ranking the state as the fifth most-popular vacation destination, behind California, Florida, New York, and Texas. All of these impressive figures are a result of one factor: New Jersey's amazing geographic diversity.

In a smaller area, the northern region of New Jersey provides the same range of contrasts. Its strikingly beautiful mountains are hosts to summer excursions and winter skiing. Crystal clear lakes grace rolling green hills and accommodate almost every water sport. Historical landmarks abound, making northern New Jersey one of the most "history conscious" areas of the country. Boasting two of the state's largest attractions—Action Park and the Meadowlands Sports Complex—the northern region caters to the participant and spectator alike. All of this is juxtaposed against one of the country's most vital and active areas, with efficient corporate complexes and quiet residential communities. Clearly, northern New Jersey's offerings are plentiful and diverse enough to satisfy the most jaded traveler. It is a microcosm of America's recreational possibilities.

Bluefishing off the New Jersey coastline proves to be both a challenging and a rewarding hobby. Photo by Bob Krist

PATRONS

The following individuals, companies and organizations have made a valuable commitment to the quality of this publication. Windsor Publications and The Commerce & Industry Association of New Jersey gratefully acknowledge their participation in *Northern New Jersey: Gateway to the World Marketplace.*

Becton Dickinson and Company*
Belgiovine Construction Group*
The Boc Group*
Citizens First National Bank*
Cole, Schotz, Bernstein, Meisel & Forman, P.A.*
CPC International Inc.*
J. Fletcher Creamer & Son, Inc.*
Ernst & Whinney*
Grant Thornton, Accountants and Management Consultants*
Hackensack Medical Center*
Jaguar Cars Inc.*
Jersey Printing and Office Supply Co., Inc.*
Loews Glenpointe Hotel*
Midlantic National Bank/North*
New Jersey Bell*
Olsten Services of Northern New Jersey*
The Port Authority of New York and New Jersey*
Rockland Electric Company*
Shanley & Fisher, P.C.*
Sheraton Meadowlands Hotel*
United Water Resources*
The Woodcliff Lake Hilton*

*Corporate Profiles in *Northern New Jersey: Gateway to the World Marketplace.* The histories of these companies and organizations appear throughout the book.

Bibliography

Albion, Robert Greenhalgh. *"Early Communications." The Story of New Jersey,* William Starr Myers, ed., pp. 47-48. New York: Lewis Historical Publishing Company, Inc., 1945.

"Big Banks Getting Bigger." Outlook '88, Part 2. *Star Ledger* (January 24, 1988): 1-22.

Bishop, Gordon. *Gems of New Jersey.* Englewood Cliffs: Prentice Hall, Inc., 1985.

Bracco, Edgar J. "Teterboro Airport: Still Making Aviation History." *New Jersey Living* (March 1987): 23-25.

Coffin, C.W. Floyd. "Bergen County—Planned Industrial Giant." *New Jersey Business.* (June 1960): 28-29.

Cramer, Jerome H. *New Jersey in the Automobile Age: A History of Transportation.* Princeton: D. Van Nostrand Company, Inc., 1964.

Cumming, Dean H. "New Jersey Malls: More than A Market Place." *Suburban New Jersey Life* (May 1973): 56-70.

Cunningham, John T. *Newark.* Newark: New Jersey Historical Society, 1966.

Cunningham, John T. *New Jersey: America's Main Road.* New York: Doubleday & Company, Inc., 1966.

Cunningham, John T. *The New Jersey Sampler.* New Jersey: The New Jersey Almanac, Inc., 1964.

Cunningham, John T. *This Is New Jersey.* New Brunswick: Rutgers University Press, 1978.

Didion, Joan. *The White Album.* New York: Simon and Schuster, 1979.

Easterbrook, Gregg. "The Revolution." *Newsweek* (January 26, 1987): 40-74.

Eisenman, Paul E. "Shopping Centers: Revolution Or Evolution." *New Jersey Business* (June 1960): 22-24.

Facaros, Dana, and Michael Pauls. *New York and the Mid-Atlantic States: A Handbook For the Independent Traveler.* Chicago: Regnery Gateway, 1982.

"The Fatal Blunder." *Quarterly Review of Literature* 2 (1944): 126.

Fogerty, Catherine M., John E. O'Connor, and Charles F. Cummings. *Bergen County: A Pictorial History.* Norfolk/Virginia Beach: The Donning Company, 1985.

Frey, Jennifer. "Rockaway Townsquare Goes Upscale." *The Daily Record.*

"Garden State Plaza: Success In the Suburbs." *New Jersey Business* (November 1958): 26-27.

"The 'Gold Coast' of Hudson still Booming." Outlook '88. *Star Ledger.* Section Ten (January 24, 1988): 90.

Graulich, William III. "Hotel Industry Rapidly Changing." New Jersey Success. Economic Forecast 1988. Vol. 8, no. 4. Hillside, N.J.: Success Publishing Co., Inc., 1987.

Hannau, Hans W. *New Jersey.* New York: Doubleday & Company, Inc.

"Harborside Readies A New Phase." *New Jersey Business.* (December 1987): 58-59.

Houstoun, Lawrence, Jr. "Pedestrian Malls—A Tale of Two Cities +." *New Jersey Municipalities.* (February 1986): 6-7, 32.

Ibert, Deborah J. "Hospitals Win Fee Hike For Nurse Pay." *The Bergen Record* (August 20, 1987): 1.

Juster, Jacqueline. "Corporate Volunteerism." *New Jersey Business* (June 1982): 37-40.

Juster, Jacqueline. "What's Happening At Our Hospitals?" *New Jersey Business* (February 1987): 15-16.

Lee, Francis Bazley. *New Jersey As A Colony And As A State.* New York: The Publishing Society of New Jersey, 1902.

Leiby, Adrian C. *The Revolutionary War In The Hackensack Valley: The Jersey Dutch and the Neutral Ground.* New Brunswick: Rutgers University Press, 1962.

Le Maire, John S. "Business Survey Of Bergen County." *New Jersey Business* (June 1960): 31-32.

Levin, William. *A Story of New Jersey Journalism.* Newark: 1928.

Lueck, Thomas J. "Jersey, Boasting Lower Costs, Leads Nation in New Offices." *The New York Times* (Tuesday, June 9, 1987).

Lueck, Thomas J. "Jersey Realty Coup by Hartz." *The New York Times* (August 8, 1987).

Mariani, Paul. *William Carlos Williams: A New World Naked.* New York: McGraw Hill Book Company, 1981.

Monroe, James A., and Andrew B. Dunham. "Slouching Toward National Health Insurance; the Unanticipated Politics of Drugs." *Bulletin of N.Y. Academy of Medicine.* Vol. 62, no. 6 (July-August 1986).

Naisbitt, John. *The Year Ahead 1986: The Powerful Trends Shaping Your Future.* New York: Warner Books Inc., 1985.

"Newport: A New City On Hudson." *New Jersey Business* (December 1987): 53-56.

Parsons, Floyd W. *New Jersey: Life, Industries & Resources of A Great State.* Newark: New Jersey State Chamber of Commerce, 1928.

Peterson, Walter Scott. *An Approach to Paterson.* New Haven: Yale University Press, 1967.

"Planned Shopping Centers." *Review of New Jersey Business* (October 1953): 8-11.

Prior, James. "The Complex World of Shopping Centers." *New Jersey Business* (July 1973): 14-22, 33.

Prior, James T. "Jersey City: Gem of the Gold Coast." *New Jersey Business* (December 1987): 35-51.

Prior, James T. "The Small Worlds Of Big Shopping Centers." *New Jersey Business* (July 1970): 21, 27, 50.

Prior, James T., ed. "Shipping Lanes Set Sail For New Jersey: Ports Newark-Elizabeth Thriving." *New Jersey Business* (November 1987): 55-60.

Riesman, David, Nathan Glazer, and Reuel Denney. *The Lonely Crowd.* New York: Doubleday & Company, Inc., 1950.

Rottenberg, Dan. "The Mall Mystique." *New Jersey Monthly* (March 1977): 31-34, 80-83.

Rust, Albert Orion. "An Analysis of the Relationship Between New Jersey Suburban Shopping Centers and Home Values In Adjacent Neighborhoods." Ph.D. dissertation. New York University, 1972.

Sachsman, David, and Warren Sloat. *The Press and the Suburbs, The Daily Newspapers of New Jersey.* Center for Urban Policy Research, Rutgers, The State University of New Jersey, 1985.

Sales & Marketing Management, Survey of Buying Power, 1986.

Sanzari, David. "Downtown Project's Success Rewards Developer's Vision." *Real Estate Forum* (February 1988): 426.

Smith, Brian. "Corporate Headquarters Grow In New Jersey." *New Jersey Business* (May 1981): 71-73, 110-111.

Somers, Anne. "N.J. Hospitals React to the 'Geriatric Imperative.'" *Hospitals* (October 20, 1986): 108.

Stockton, Frank R. *Stones of New Jersey.* New Brunswick: Rutgers University Press, 1961.

"13 Artists Awarded $15,000 State Grants." Section 11, N.J. Weekly Sun. *The New York Times* (August 30, 1987): 23.

Whitlow, Joan. "Jersey Hospital Officials Review Rating as Nation's Health-Care Bargain." *Star Ledger,* year-end issue, 1986.

Whittemore, Reed. *William Carlos Williams, Poet From Jersey.* Boston: Houghton Mifflin Company, 1975.

Williams, William Carlos. *Selected Poems.* New York: New Directions Publishing Corporation, 1968.

Index